中 外 物 理 学 精 品 书 系

中外物理学精品书系

高瞻系列·11

Applied Analysis for Engineering Sciences
工程科学中的应用分析

唐少强 编著

北京大学出版社
PEKING UNIVERSITY PRESS

图书在版编目(CIP)数据

工程科学中的应用分析 = Applied Analysis for Engineering Sciences：英文 / 唐少强编著. — 北京：北京大学出版社，2016.3
（中外物理学精品书系）
ISBN 978-7-301-26761-5

Ⅰ.①工… Ⅱ.①唐… Ⅲ.①工程技术－研究－英文 Ⅳ.①TB1

中国版本图书馆CIP数据核字(2016)第010064号

书　　名	Applied Analysis for Engineering Sciences（工程科学中的应用分析）
著作责任者	唐少强　编著
责任编辑	刘啸
标准书号	ISBN 978-7-301-26761-5
出版发行	北京大学出版社
地　　址	北京市海淀区成府路205号　100871
网　　址	http://www.pup.cn
电子信箱	zpup@pup.cn
新浪微博	@北京大学出版社
电　　话	邮购部 62752015　发行部 62750672　编辑部 62752021
印刷者	北京中科印刷有限公司
经销者	新华书店
	730毫米×980毫米　16开本　10印张　172千字
	2016年3月第1版　2016年3月第1次印刷
定　　价	30.00元

未经许可，不得以任何方式复制或抄袭本书之部分或全部内容。
版权所有，侵权必究
举报电话：010-62752024　电子信箱：fd@pup.pku.edu.cn
图书如有印装质量问题，请与出版部联系，电话：010-62756370

"中外物理学精品书系"
编委会

主　　任：王恩哥
副主任：夏建白
编　　委：(按姓氏笔画排序,标*号者为执行编委)

王力军	王孝群	王　牧	王鼎盛	石　兢
田光善	冯世平	邢定钰	朱邦芬	朱　星
向　涛	刘　川*	许宁生	许京军	张　酣*
张富春	陈志坚*	林海青	欧阳钟灿	周月梅*
郑春开*	赵光达	聂玉昕	徐仁新*	郭　卫*
资　剑	龚旗煌	崔　田	阎守胜	谢心澄
解士杰	解思深	潘建伟		

秘　　书：陈小红

序　言

物理学是研究物质、能量以及它们之间相互作用的科学。她不仅是化学、生命、材料、信息、能源和环境等相关学科的基础，同时还是许多新兴学科和交叉学科的前沿。在科技发展日新月异和国际竞争日趋激烈的今天，物理学不仅囿于基础科学和技术应用研究的范畴，而且在社会发展与人类进步的历史进程中发挥着越来越关键的作用。

我们欣喜地看到，改革开放三十多年来，随着中国政治、经济、教育、文化等领域各项事业的持续稳定发展，我国物理学取得了跨越式的进步，做出了很多为世界瞩目的研究成果。今日的中国物理正在经历一个历史上少有的黄金时代。

在我国物理学科快速发展的背景下，近年来物理学相关书籍也呈现百花齐放的良好态势，在知识传承、学术交流、人才培养等方面发挥着无可替代的作用。从另一方面看，尽管国内各出版社相继推出了一些质量很高的物理教材和图书，但系统总结物理学各门类知识和发展，深入浅出地介绍其与现代科学技术之间的渊源，并针对不同层次的读者提供有价值的教材和研究参考，仍是我国科学传播与出版界面临的一个极富挑战性的课题。

为有力推动我国物理学研究、加快相关学科的建设与发展，特别是展现近年来中国物理学者的研究水平和成果，北京大学出版社在国家出版基金的支持下推出了"中外物理学精品书系"，试图对以上难题进行大胆的尝试和探索。该书系编委会集结了数十位来自内地和香港顶尖高校及科研院所的知名专家学者。他们都是目前该领域十分活跃的专家，确保了整套丛书的权威性和前瞻性。

这套书系内容丰富，涵盖面广，可读性强，其中既有对我国传统物理学发展的梳理和总结，也有对正在蓬勃发展的物理学前沿的全面展示；既引进和介绍了世界物理学研究的发展动态，也面向国际主流领域传播中国物理的优秀专著。可以说，"中外物理学精品书系"力图完整呈现近现代世界和中国物理科学发展的全貌，是一部目前国内为数不多的兼具学术价值和阅读乐趣的经典物理丛书。

"中外物理学精品书系"另一个突出特点是，在把西方物理的精华要义"请进来"的同时，也将我国近现代物理的优秀成果"送出去"。物理学科在世界范围内的重要性不言而喻，引进和翻译世界物理的经典著作和前沿动态，可以满足当前国内物理教学和科研工作的迫切需求。另一方面，改革开放几十年来，我国的物理学研究取得了长足发展，一大批具有较高学术价值的著作相继问世。这套丛书首次将一些中国物理学者的优秀论著以英文版的形式直接推向国际相关研究的主流领域，使世界对中国物理学的过去和现状有更多的深入了解，不仅充分展示出中国物理学研究和积累的"硬实力"，也向世界主动传播我国科技文化领域不断创新的"软实力"，对全面提升中国科学、教育和文化领域的国际形象起到重要的促进作用。

值得一提的是，"中外物理学精品书系"还对中国近现代物理学科的经典著作进行了全面收录。20世纪以来，中国物理界诞生了很多经典作品，但当时大都分散出版，如今很多代表性的作品已经淹没在浩瀚的图书海洋中，读者们对这些论著也都是"只闻其声，未见其真"。该书系的编者们在这方面下了很大工夫，对中国物理学科不同时期、不同分支的经典著作进行了系统的整理和收录。这项工作具有非常重要的学术意义和社会价值，不仅可以很好地保护和传承我国物理学的经典文献，充分发挥其应有的传世育人的作用，更能使广大物理学人和青年学子切身体会我国物理学研究的发展脉络和优良传统，真正领悟到老一辈科学家严谨求实、追求卓越、博大精深的治学之美。

温家宝总理在2006年中国科学技术大会上指出，"加强基础研究是提升国家创新能力、积累智力资本的重要途径，是我国跻身世界科技强国的必要条件"。中国的发展在于创新，而基础研究正是一切创新的根本和源泉。我相信，这套"中外物理学精品书系"的出版，不仅可以使所有热爱和研究物理学的人们从中获取思维的启迪、智力的挑战和阅读的乐趣，也将进一步推动其他相关基础科学更好更快地发展，为我国今后的科技创新和社会进步做出应有的贡献。

"中外物理学精品书系"编委会　主任
中国科学院院士，北京大学教授
王恩哥
2010年5月于燕园

Contents

Preface · iii

Chapter 1 Qualitative Theory for ODE Systems · · · · · · · · · · · · · 1
   1.1 Basic notions · 1
   1.2 Local existence · 3
      1.2.1 Normed spaces and fixed point theorem · · · · · · · · · · 4
      1.2.2 Applications to ODE system and linear algebraic system · · · 11
   1.3 Critical point · 14
   1.4 Plane analysis for the Duffing equation · · · · · · · · · · · · · · 18
   1.5 Homoclinic orbit and limit cycle · · · · · · · · · · · · · · · · · · 24
   1.6 Stability and Lyapunov function · · · · · · · · · · · · · · · · · · 29
   1.7 Bifurcation · 33
   1.8 Chaos: Lorenz equations and logistic map · · · · · · · · · · · · · 38

Chapter 2 Reaction-Diffusion Systems · · · · · · · · · · · · · · · · · · 50
   2.1 Introduction: BVP and IBVP, equilibrium · · · · · · · · · · · · · 50
   2.2 Dispersion relation, linear and nonlinear stability · · · · · · · · · 57
   2.3 Invariant domain · 60
   2.4 Perturbation method · 63
   2.5 Traveling waves · 69
   2.6 Burgers' equation and Cole-Hopf transform · · · · · · · · · · · · 72
   2.7 Evolutionary Duffing equation · · · · · · · · · · · · · · · · · · · 74

Chapter 3 Elliptic Equations · 86
   3.1 Sobolev spaces · 86
   3.2 Variational formulation of second-order elliptic equations · · · · · 88
   3.3 Neumann boundary value problem · · · · · · · · · · · · · · · · · 93

Chapter 4 Hyperbolic Conservation Laws · · · · · · · · · · · · · · · · 95
   4.1 Linear advection equation, characteristics method · · · · · · · · · 95
   4.2 Nonlinear hyperbolic equations · · · · · · · · · · · · · · · · · · · 97
   4.3 Discontinuities in inviscid Burgers' equation · · · · · · · · · · · · 101

4.4 Elementary waves in inviscid Burgers' equation · · · · · · · · · · 103
4.5 Wave interactions in inviscid Burgers' equation · · · · · · · · · · 107
4.6 Elementary waves in a polytropic gas · · · · · · · · · · · · · · · 114
4.7 Riemann problem in a polytropic gas · · · · · · · · · · · · · · · 121
4.8 Elementary waves in a polytropic ideal gas · · · · · · · · · · · · 126
4.9 Soliton and inverse scattering transform · · · · · · · · · · · · · · 128

Index · 144

Preface

Throughout the history of civilization, mathematics has served as one of the major tools to analyze real world applications. In turn, through these applications it has been developed and expanded considerably. Moreover, mathematics helps establishing and consolidating the belief in eternal and exact truth, and hence the trust on sciences.

Since the invention of Calculus by Newton and Leibniz in the seventeenth century, mathematics has been overwhelmingly successful in almost every branch of sciences. It is instrumental for scientists and engineers to think, to work and to communicate.

We recall that Calculus is built fundamentally upon the definition of the real numbers. This definition naturally leads to the notion of limit. Two special and most useful limits are the derivative and the integral of a function. Most physical theories are described in terms of differential equations. Modern physics essentially started from Newton's theory of motions, and Newton's second law is a paradigm. The electro-magnetic theory is essentially the studies on the Maxwell equations. The theory of general relativity explores the Einstein equation, and the quantum mechanics uses the Schrödinger equation or the Wigner equation. We discuss the Lagrangian or Hamiltonian systems in mechanics, the biharmonic equation in elasticity, and the Navier-Stokes equations in fluid, etc.

The invention of electronic computers changed fundamentally the way for scientific research. Though we may not obtain analytical solution to a complex system in general, computer allows us to find the solution to a set of continuous differential equations in a discrete manner. Under certain circumstances, we even do not need to go to the continuous form. For instance, a fully discrete binomial algorithm may be used to compute the price of an option. We point out that instead of becoming a substitute for the continuous analysis, scientific computing reaches its best efficiency in real world applications only when we have a good understanding of the physics, the continuous modeling and analysis, the numerical algorithm, and the computer code.

This book is an outcome of an advanced course, conducted in English, for graduate students and senior undergraduate students at Department of Mechanics of Peking University. This course has been offered roughly every other year since 1998. We set forth the following objectives.

- To show some modern (1900-1990?) mathematical methods that are widely used in engineering sciences, nonlinear mechanics and other physical sciences.
- To help initiating research activities, namely, to boost ideas, to formulate the problem, and to explore the mathematics.
- To help bridging the gap between the mathematical tools and the physical understandings taught in other undergraduate courses.

A major ingredient of this course is nonlinearity. As is well known, superposition is the feature that distinguishes linear and nonlinear systems.

In linear algebra, we have

$$Ax_1 = y_1, \quad Ax_2 = y_2 \quad \Rightarrow \quad A(x_1 + x_2) = y_1 + y_2.$$

The differential operator and integral operator are also linear.

$$\frac{d}{dx}(\alpha f(x) + \beta g(x)) = \alpha \frac{df}{dx} + \beta \frac{dg}{dx},$$

$$\int (\alpha f(x) + \beta g(x)) \, dx = \alpha \int f(x) dx + \beta \int g(x) dx.$$

Similarly, the Fourier transform and the Laplace transform are linear operators.

Superposition also applies to linear differential equations. For example, if both $x_1(t)$ and $x_2(t)$ are solutions to the equation $ax'' + bx' + cx = 0$, so is $\alpha x_1(t) + \beta x_2(t)$ for any constant α and β. As a matter of fact, the general solution is $x(t) = C_1 e^{\lambda_1 t} + C_2 e^{\lambda_2 t}$, where λ_1 and λ_2 are the roots to the quadratic equation $a\lambda^2 + b\lambda + c = 0$.

For instance, we consider the following RCL circuit in Fig. 1. A resistor obeys Ohm's law $V_R = RI$, while an inductor and a capacitor satisfy $V_L = L\frac{dI}{dt}$ and $I =$

$C\frac{dV_C}{dt}$, respectively. Kirchhoff's law gives rise to an integral-differential equation.

$$V = RI + L\frac{dI}{dt} + \frac{1}{C}\int_0^t I(s)ds + V_C(0).$$

We differentiate it once to obtain

$$L\frac{d^2 I}{dt^2} + R\frac{dI}{dt} + \frac{1}{C}I = 0.$$

If V varies along with time, the righthand side does not vanish. This circuit may generate electro-magnetic waves of a certain frequency.

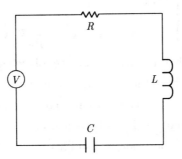

Figure 1 RCL circuit.

As Einstein pointed out, the Laws of Nature cannot be linear. Linear system is usually a special case or a simplified version, therefore incomplete. Lots of important features of the real world can only be explained under the framework of nonlinear systems. Besides, linear problems are relatively simple, and mathematical tools we have learned before are fairly competent to handle them. We head for challenges and excitements through studies of nonlinear problems.

For an example of nonlinear system, we consider an oversimplified mechanical system consisted of the sun with mass M and the earth with mass m in Fig. 2. Let their positions be y and x, respectively. Newton's second law and the universal gravitation theory lead to the following coupled system,

$$\begin{cases} m\dfrac{d^2 x}{dt^2} = \dfrac{GmM}{|x-y|^2}, \\ M\dfrac{d^2 y}{dt^2} = -\dfrac{GmM}{|x-y|^2}. \end{cases}$$

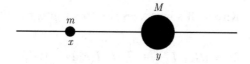

Figure 2   A one-dimensional two-body system of the sun and the earth.

For another example, the Navier-Stokes equations for an incompressible fluid are as follows. Let **v** be the velocity, p the pressure, and Re the Reynolds number,

$$\begin{cases} \nabla \cdot \mathbf{v} = 0, \\ \mathbf{v}_t + (\mathbf{v}\nabla) \cdot \mathbf{v} + \nabla p = \frac{1}{\text{Re}} \nabla^2 \mathbf{v}. \end{cases}$$

In this course, we shall mainly discuss the following topics.

In Chapter 1, we expose the qualitative theory for ODE systems (4 weeks of teaching). We start with some basic notions. Then we present a basic fixed point theory from functional analysis. This allows us to establish existence results for an ODE system. A further application is also illustrated, namely, iteration methods for solving a linear algebraic system. To understand qualitatively an ODE system, we analyze its critical points. For a second order ODE, the so-called plane analysis may provide substantial understanding. For a general system, there are not as many powerful tools. Stability analysis via the Lyapunov function is an exception. When there is a controlling parameter in a system, bifurcation may occur. We conclude this chapter by an exhibition of chaos in the Lorenz system and the logistic map.

For partial differential equations, we first study reaction-diffusion systems (3 weeks of teaching). We set up BVP (boundary-value problem) and IBVP (initial-boundary-value problem), and then show a simple example of instability at equilibrium. For a linearized problem, its dispersion relation gives a primary linear stability result. For nonlinear systems, an invariant domain approach sometimes works. This is a geometrical way to get *a priori* estimate. For a special example of nonlinear system, we illustrate a perturbation method for its steady states. Next, traveling wave analysis reduces a PDE system to an ODE system, and usually provides explanation to some wave behaviours of the PDE system. Only for very exceptional cases, a nonlinear PDE may be transformed to a linear one, e.g.,

Burgers' equation by the Cole-Hopf transform. We further illustrate a combination of theoretical and numerical investigations in an example of reaction-diffusion equation, namely, the evolutionary Duffing equation.

In Chapter 3, we discuss elliptic equations (2 weeks of teaching). the main topic is to introduce some basic ideas in the modern theories of partial differential equations. We start with generalized functions and weak derivatives, and introduce briefly the Sobolev spaces, and state the embedding theorem. Weak formulations and minimization procedure are used to establish existence results.

Chapter 4 is devoted to hyperbolic conservation laws (5 weeks of teaching). The most distinct feature of this type of PDE's lies in the inevitable appearance of discontinuities, regardless of smooth initial data. We show shock formation in inviscid Burgers' equation, by a characteristics approach. Then taking the Euler equations for polytropic gas as an example, we discuss the elementary waves, which include shock waves via vanishing viscosity approach, and rarefaction waves via self-similarity solution approach. For a general Riemann problem of gas dynamics, the unique composition of these elementary waves gives the solution, which is a weak one by construction. We further discuss solitons in the KdV equation, for which a brilliant theory of inverse scattering transform is sketched.

As this book is only an introduction of qualitative theories for ODE and PDE systems, further readings are suggested.

1. Smoller J. Shock Waves and Reaction-diffusion Equations. Springer, 1999.
2. Grindrod P. Patterns and Waves. Claredon, 1991.
3. Whitham G B. Linear and Nonlinear Waves. John Wiley & Sons, 1974.
4. Wang L, Wang M Q. Qualitative Analysis for Nonlinear Ordinary Differential Equations (in Chinese). Harbin Institute of Technology Press, 1987.
5. Huang Y N. Lecture Notes on Nonlinear Dynamics (in Chinese). Peking University Press, 2010.
6. Ding T R, Li C Z. A Course on Ordinary Differential Equations (in Chinese). Higher Education Press, 2004.
7. Ye Q X, Li Z Y. Introduction to Reaction-Diffusion Equations (in Chinese). Science Press, 1994.
8. Braess D. Finite Elements. Cambridge University Press, 2001.

Acknowledgements

I appreciate all our excellent students for their help over the years. They kindly offer me the motivation to write this book. Among them, Huanlong Li, Xiangming Xiong, Ziwei Yang, Chunbo Wang, Jianchun Wang, and Shengkai Wang have assisted me in the preparation of the lecture notes. I would like to thank Professor Zhaoxuan Zhu and Professor Yongnian Huang, who offered similar courses in Department of Mechanics, Peking University, and kindly shared their ideas with me. The course and book have been supported partially by the National Bilingual Course Supporting Project of Ministry of Education, and a project of the Peking University Press.

Chapter 1 Qualitative Theory for ODE Systems

1.1 Basic notions

For many applications, we describe a system using only one independent variable. A dependent quantity is expressed as a function of this independent variable. An ordinary differential equation (ODE) is an equation that contains an unknown function, called a state variable, together with its derivatives with respect to the single independent variable. For historical reasons, the independent variable is typically denoted as t, representing time. Depending on the applications, actually t may mean some other quantities, such as temperature, height, etc. The order of the highest derivative in an ODE is its order.

An ODE system, also called a dynamical system, is a set of ODE's. Typically each equation in this system is of first order. The order of the ODE system is the number of first order equations in the system.

A high-order ODE can always be recast to an ODE system of the same order. For instance,

$$x'' + xx' + x(1-x) + f(t) = 0 \tag{1.1}$$

can be rewritten as

$$\begin{cases} x' = y, \\ y' = -[xy + x(1-x) + f(t)]. \end{cases} \tag{1.2}$$

Therefore, a general ODE system reads

$$x' = f(t, x), \quad \text{with} \quad x = (x_1, \cdots, x_n)^T \in \mathbb{R}^n. \tag{1.3}$$

The ODE system is autonomous if the righthand side depends only on the state variable, that is, $f(x, t) = f(x)$. A non-autonomous system can be trivially reshaped to an autonomous one. In fact, if we take $y = (t, x_1, \cdots, x_n)^T$, then we obtain

$$y' = \begin{pmatrix} 1 \\ f(y) \end{pmatrix}. \tag{1.4}$$

Through this procedure, the order of the system rises by one.

In this course, we are not concerned with a particular solution to an ODE system. Instead, we take a global view of all the solutions to a system, and for the aforementioned reason, an autonomous ODE system. These solutions form a family of (vector) functions. This family is the object for the qualitative theory.

If one such function $x(t)$ satisfies $x(t_0) = x_0$ for a certain time t_0, then we call it an orbit, or a trajectory passing through the point (t_0, x_0). These names reflect that we take a geometrical view. Sometimes we use the notation $x = \phi(t; t_0, x_0)$ to identify the orbit. In contrast, in the previous ODE course, one regards (t_0, x_0) as an initial point, and usually considers the solution $x = x(t)$ only for $t \geq t_0$. The geometrical name for this part of the solution is the positive semi-orbit, denoted by $x = \phi^+(t; t_0, x_0)$. Meanwhile, the solution for $t \in (-\infty, t_0]$ is called as the negative semi-orbit, and denoted as $x = \phi^-(t; t_0, x_0)$. Under such a geometrical view, we regard the function $x(t)$ equivalent to a curve in the space $\mathbb{R}^n$. This space is called a phase space, or a phase plane if $n = 2$.

We remark that sometimes one also specifies boundary data, namely n algebraic equations for n quantities selected from $x_1(a), \cdots, x_n(a), x_1(b), \cdots, x_n(b)$, when one looks for a solution in $t \in [a, b]$.

For an autonomous system, a translation in time is invariant. More precisely, if $x = \phi(t; 0, x_0)$ solves the system with initial data $x(0) = x_0$, then $x = \phi(t+t_0; t_0, x_0)$ solves the system with initial data $x(t_0) = x_0$. Therefore, it suffices to study the problem with initial data at one selected time, which is usually chosen as $t_0 = 0$.

We recall that the existence, uniqueness and continuous dependency hold for quite general cases, e.g., when the source term (righthand side) is continuous. Existence will be discussed later by means of a fixed point theory.

Qualitative theory is concerned with the global structure of trajectories in the phase space, instead of a particular solution for certain given initial data.

At each given point x, the source term introduces a vector $f(x)$ in the phase space. The direction of this vector determines the direction of the trajectory, and the absolute value determines how fast a solution takes to go through this point. We may imagine that there is a particle moving along the trajectory according to the vector field (velocity field). See Fig. 1.1.

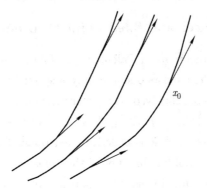

Figure 1.1   Trajectories, vector field and phase space (plane).

We notice that at a point x where $f(x)$ vanishes, the previous statement becomes meaningless. This leads to the notion of a critical point (equilibrium point, singular point, stationary point, etc.), which turns out to be crucial in later discussions.

A point x is a critical point where $f(x) = 0$. It is a regular point if f is finite and non-zero. Two trajectories may intersect only at a critical point in the phase space. This can be proved by the uniqueness of solution to the following ODE system in the neighborhood of a regular point x^*. Assuming that $f_1(x) \neq 0$, we have

$$\frac{dx_2}{dx_1} = \frac{f_2(x)}{f_1(x)}, \cdots, \frac{dx_n}{dx_1} = \frac{f_n(x)}{f_1(x)}, \quad (x_2(x_1^*), \cdots, x_n(x_1^*)) = (x_2^*, \cdots, x_n^*)). \quad (1.5)$$

Furthermore, a trajectory usually starts from/ends at a critical point or infinity, or forms a closed orbit. Under certain circumstances, a chaotic orbit may appear.

In the subsequent sections, we shall discuss the local existence for ODE systems, using a fixed point theory. Then we shall perform detailed analysis in the vicinity of a critical point.

1.2   Local existence

The local existence for an ODE system may be proven through a Picard iteration procedure. We present here a systematic approach, using a fixed point theory from functional analysis.

1.2.1 Normed spaces and fixed point theorem

One of the fixed point theorems in Calculus is as follows. If a mapping $f : \mathbb{R} \to \mathbb{R}$ is a contraction, then there exists one and only one fixed point.

Here are some concepts involved. By a contraction, we mean that $\exists \alpha \in [0,1)$, $\forall x, y \in \mathbb{R}$, it holds that $|f(x) - f(y)| \leqslant \alpha |x - y|$. A contractive function, is necessarily continuous. A point $x \in \mathbb{R}$ is called a fixed point if $f(x) = x$, i.e., f maps this point to itself. The proof for this type of theorems uses an iteration procedure.

Our interest goes beyond this simple situation. We intend to develop tools to analyze differential equations. For this purpose, we work in function spaces, and define contractive operators.

Space, similar to set, is a primitive concept. It consists of elements/points. In a space, different points are not related to each other in general, except that they are in the same space. A vector space is a space X with points (called vectors now) $x, y, \cdots$, in which the vector addition and the multiplication by scalars are defined. These two operations provide the algebraic structure for a space.

More precisely, vector addition is a binary operation that puts any two vectors x, y to another vector $z = x + y \in X$, called the sum of x and y. It is Abelian and associative, i.e., for any vectors $x, y, z \in X$, one has

$$x + y = y + x, \quad \text{(Abelian)} \tag{1.6}$$

$$(x + y) + z = x + (y + z). \quad \text{(associative)} \tag{1.7}$$

Moreover, there exists a zero vector $0 \in X$, which satisfies $0 + x = x, \forall x \in X$. For each $x \in X$, there is a unique $-x \in X$, and it holds that $x + (-x) = 0$.

Multiplication of a vector $x \in X$ by a scalar $\lambda \in \mathbb{R}$ defines a product $\lambda x \in X$. It satisfies the following rules,

$$\alpha(\beta x) = (\alpha \beta) x, \quad \forall \alpha, \beta \in \mathbb{R}, \forall x \in X; \tag{1.8}$$

$$1 \cdot x = x, \quad \forall x \in X; \tag{1.9}$$

$$\alpha(x + y) = \alpha x + \alpha y, \quad \forall \alpha \in \mathbb{R}, x, y \in X; \tag{1.10}$$

$$(\alpha + \beta) x = \alpha x + \beta x, \quad \forall \alpha, \beta \in \mathbb{R}, x \in X. \tag{1.11}$$

Same as in linear algebra, we call a given finite set $M = \{x_1, \cdots, x_n\}$ linearly independent if $\sum_{i=1}^{n} \lambda_i x_i = 0$ implies that $\lambda_i = 0, i = 1, \cdots, n$. We call an infinite set M to be linearly independent if every finite subset of M is linearly independent.

A vector space X is r-dimensional, if every $(r+1)$ vectors in this space are linearly dependent, while there exist r vectors which are linearly independent. An infinite dimensional space refers to a vector space that does not have a finite dimension, i.e., there exist n vectors which are linearly independent for any $n \in \mathbb{N}$.

We illustrate this with an example of $\mathbb{R}^2$. We know that the vectors $(1,0)$ and $(0,1)$ are linearly independent. On the other hand, any three vectors must be linearly dependent. In fact, if the first two vectors are linearly independent, then the third one can be expressed as a linear combination of the first two. See Fig. 1.2.

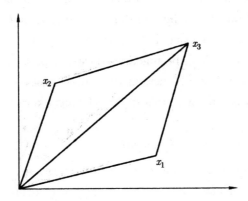

Figure 1.2   Linear dependency of three vectors in $\mathbb{R}^2$.

Now we introduce a geometric structure to a vector space. A norm is a function on a vector space X. That is, for each point $x \in X$, we denote its norm as $\|x\| \in \mathbb{R}$. This norm function satisfies the following four axioms.

1. $\|x\| \geqslant 0, \ \forall x \in X$;
2. $\|x\| = 0 \Leftrightarrow x = 0$;
3. $\|\alpha x\| = |\alpha| \, \|x\|, \ \forall \alpha \in \mathbb{R}, \forall x \in X$;
4. $\|x + y\| \leqslant \|x\| + \|y\|, \ \forall x, y \in X$. (triangular inequality)

The distance between two vectors x and y is defined as $\|x - y\|$. We notice that when distance is defined, the definition of limit follows.

A normed space is a vector space with a norm. We remark that a vector space may take different norms, and the resulting normed spaces are different. In another word, only when both the algebraic structure and the geometric structure are the same, may we identify two normed spaces.

The Euclidean space $\mathbb{R}^n$ with $\|x\| = \left(\sum_{i=1}^n x_i^2\right)^{1/2}$ is a simple example. One may easily verify that this defines a normed space. Historically, people sometimes write $\mathbb{E}^n$ to emphasize that the norm is the Euclidean one. In the mean time, $\|x\|_{\max} = \max_{1 \leqslant i \leqslant n} |x_i|$ also satisfies the axioms and defines another norm on $\mathbb{R}^n$. We remark that in case of finite dimensional spaces, it may be proved that different norms are equivalent. Moreover, to certain extent, an n-dimensional normed space is essentially $\mathbb{R}^n$. However, for infinite dimensional spaces, different norms make two normed spaces with the same underlying vector space different.

Now we are ready to describe function spaces. In particular, we consider the space $C[a,b]$, which contains all functions that are continuous over a closed interval $[a,b]$. Each function is a point in this space. The algebraic operations are defined as usual by

$$(f+g)(t) = f(t) + g(t), \quad (\alpha f)(t) = \alpha f(t). \tag{1.12}$$

This defines an infinite dimensional vector space. For instance, we consider a specific example of $C[0,1]$. For any given integer n, $f_n(t) = t^n$ defines a point in $C[0,1]$. Moreover, $f_1, \cdots, f_n$ are linearly independent.

For the geometric structure, we define

$$\|f\| = \max_{t \in [a,b]} |f(t)|. \tag{1.13}$$

Because $f \in C[a,b]$, this maximum is attained, and the norm is well defined. All four axioms are readily verified. This maximum norm is analogous to $\|x\|_{\max}$ in $\mathbb{R}^n$.

Meanwhile, we may define an L^2 norm

$$\|f\|_2 = \left[\int_a^b |f(t)|^2 dt\right]^{1/2}. \tag{1.14}$$

It also satisfies all axioms, and resembles the Euclidean norm for $\mathbb{R}^n$. However, we claim that under these two norms, the function spaces are different, from the perspective of completeness.

A point $x \in X$ is the limit of a sequence $\{x_n\} \subset X$, if $\lim_{n\to\infty} \|x_n - x\| = 0$. In another word, we call $\lim_{n\to\infty} x_n = x$ if $\forall \varepsilon > 0, \exists N \in \mathbb{N}$, such that $\|x_n - x\| < \varepsilon$ provided $n > N$.

In Calculus, we have learned that a Cauchy sequence or fundamental sequence is a sequence $\{x_n\} \subset X$, for which $\forall \varepsilon > 0, \exists N \in \mathbb{N}$, such that $\|x_n - x_m\| < \varepsilon$ provided $n, m > N$. Moreover, it is proved that a Cauchy sequence is exactly a convergent sequence. That is, a Cauchy sequence converges; and a convergent sequence must be a Cauchy sequence.

In a normed space X, a convergent sequence must be a Cauchy sequence, by virtue of the triangular inequality. More precisely, if $\lim_{n\to\infty} x_n = x$, then $\forall \varepsilon > 0, \exists N \in \mathbb{N}$, such that $\|x_n - x\| < \varepsilon$ provided $n > N$. Therefore, we have

$$\|x_n - x_m\| \leqslant \|x_n - x\| + \|x_m - x\| < 2\varepsilon, \text{ for } n, m > N. \tag{1.15}$$

However, the other direction of deduction does not hold in general. If it holds, we call X a complete normed space, or a Banach space.

In particular, $\mathbb{R}^n$ is a Banach space. For $C[a, b]$, it turns out that different norms lead to different answers.

If we take the norm $\|f\| = \max_{t \in [a,b]} |f(t)|$, we claim that the space $C[a, b]$ is complete.

Assume that there is a Cauchy sequence $\{f_n\} \subset C[a, b]$. Then $\forall \varepsilon > 0$, there exists $N \in \mathbb{N}$, such that $\|f_m - f_n\| < \varepsilon$, provided $m, n > N$.

This implies that at each fixed $t \in [a, b]$, $|f_m(t) - f_n(t)| < \varepsilon$. Therefore, a real number sequence $\{f_n(t)\} \subset \mathbb{R}$ is Cauchy. Since $\mathbb{R}$ is complete, this sequence is convergent, and we denote the limit as

$$f_0(t) \equiv \lim_{n\to\infty} f_n(t). \tag{1.16}$$

When $t \in [a, b]$ varies, a function $f_0(t)$ is obtained.

If $f_0(t) \in C[a, b]$, we take the limit $n \to \infty$ in $|f_m(t) - f_n(t)| < \varepsilon$, and obtain

$$\|f_m(t) - f_0(t)\| \leqslant \varepsilon, \text{ for } m > N. \tag{1.17}$$

This proves that under the maximum norm in $C[a,b]$, it holds $\lim\limits_{n\to\infty} f_n = f_0$.

Finally, we verify that $f_0(t)$ lies in $C[a,b]$.

In fact, f_{N+1} is continuous in the closed interval $[a,b]$, so it is uniformly continuous. That is, for the above $\varepsilon > 0$, $N \in \mathbb{N}$, and any $t_1, t_2 \in [a,b]$, there exists $\delta > 0$, such that

$$|f_{N+1}(t_1) - f_{N+1}(t_2)| < \varepsilon, \text{ provided } |t_1 - t_2| < \delta. \tag{1.18}$$

This leads to

$$|f_0(t_1) - f_0(t_2)|$$
$$\leqslant |f_0(t_1) - f_{N+1}(t_1)| + |f_{N+1}(t_1) - f_{N+1}(t_2)| + |f_{N+1}(t_2) - f_0(t_2)|$$
$$< 3\varepsilon.$$

We conclude that $C[a,b]$ with the aforementioned norm is a Banach space.

In contrast, if we take the L^2 norm, $C[a,b]$ is no longer complete. To see this, we construct a sequence of continuous functions,

$$f_n(t) = \begin{cases} 1, & \text{if } t - (a+b)/2 < -1/n, \\ -n\left[t - (a+b)/2\right], & \text{if } |t - (a+b)/2| \leqslant 1/n, \\ -1, & \text{if } t - (a+b)/2 > 1/n. \end{cases} \tag{1.19}$$

This is a Cauchy sequence, because

$$\|f_n - f_m\|_2^2 = \int_a^b (f_n(t) - f_m(t))^2 dt = \frac{2(m-n)^2}{3mn \max\{m,n\}} \leqslant \frac{2}{3N}, \text{ if } m,n > N. \tag{1.20}$$

However, the pointwise limit defines a function

$$f_0(t) = \begin{cases} 1, & \text{if } t < (a+b)/2, \\ 0, & \text{if } t = (a+b)/2, \\ -1, & \text{if } t > (a+b)/2. \end{cases} \tag{1.21}$$

It is discontinuous. This shows that $C[a,b]$ is not complete under the L^2 norm. See Fig. 1.3.

Besides the completeness for a normed space, a fixed point theory works for a mapping, which is now called an operator. In fact, a function is a mapping from $\mathbb{R}$

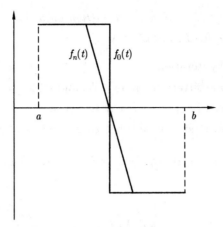

Figure 1.3 Limit of a function sequence in $C[a,b]$ with the L^2 norm.

(or its subspace) to $\mathbb{R}$. Similarly, an operator is a mapping from a normed space X to another normed space Y,

$$T : X \to Y$$
$$x \mapsto y = Tx. \tag{1.22}$$

A simple example is a linear transform, which is represented by a matrix,

$$T : \mathbb{R}^m \to \mathbb{R}^n$$
$$x \mapsto y = Ax, \tag{1.23}$$

where A is an $n \times m$ matrix.

In particular, if $Y = \mathbb{R}$, then $f : X \to \mathbb{R}$ is a functional. For special cases when X is a function space, then T is actually a function of functions. An example is the definite integral operator,

$$f : C[a,b] \to \mathbb{R}$$
$$x(t) \mapsto \int_a^b x(t) \mathrm{d}t. \tag{1.24}$$

Consider an operator from a normed space X to itself. It is contractive if $\exists \alpha \in (0,1)$, such that $\|Tx - Ty\| \leqslant \alpha \|x - y\|$ for $\forall x, y \in X$. A fixed point of T is $x \in X$ such that $Tx = x$.

The Banach fixed point theorem is as follows.

Theorem 1.2.1 *If $T : X \to X$ is a contraction on a Banach space $X \neq \Phi$, then there exists exactly one fixed point $x^* \in X$.*

The proof is done by iteration.

Step 1. We take an arbitrary point $x_0 \in X$, and let $x_n = Tx_{n-1}$ for $n \in \mathbb{N}$.

Step 2. We show that $\{x_n\}$ is a Cauchy sequence.

Because T is a contraction, there is $\alpha \in (0, 1)$, and

$$\|x_{n+1} - x_n\| \leqslant \alpha \|x_n - x_{n-1}\| \leqslant \cdots \leqslant \alpha^n \|Tx_0 - x_0\|. \tag{1.25}$$

This yields

$$\|x_{n+p} - x_n\| \leqslant (\alpha^{n+p-1} + \cdots + \alpha^n) \|Tx_0 - x_0\| \leqslant \frac{\alpha^n}{1-\alpha} \|Tx_0 - x_0\|. \tag{1.26}$$

So, $\{x_n\}$ is a Cauchy sequence.

Step 3. We find the fixed point. The space X is complete, therefore $\{x_n\}$ converges. Denote $\lim_{n\to\infty} x_n = x^*$.

Step 4. We verify that x^* is a fixed point.

$$\|Tx^* - x^*\| \leqslant \|Tx^* - Tx_n\| + \|x_{n+1} - x^*\| \leqslant \alpha \|x^* - x_n\| + \|x_{n+1} - x^*\| \to 0. \tag{1.27}$$

This shows that $Tx^* = x^*$.

Step 5. Finally we show the uniqueness. Assume that there is another fixed point x'. Then it holds that $\|Tx^* - Tx'\| = \|x^* - x'\|$. This contradicts with the fact that T is a contraction.

From this proof, we already have an error estimate

$$\|x_n - x^*\| \leqslant \frac{\alpha^n}{1-\alpha} \|Tx_0 - x_0\|. \tag{1.28}$$

The theorem actually holds for a closed subspace of a Banach space. Let X be a Banach space equipped with a norm $\|\cdot\|$, and $Y \subset X$ is a closed subspace. By a subspace, we mean that Y is a subset of X. Y inherits from X the norm hence the geometrical structure. By a closed subspace we mean that the limit of any convergent sequence $\{y_n\}_{n=1}^\infty \subset Y$ (converge in X) converges to a point in Y. Furthermore, a Cauchy sequence $\{y_i\} \subset Y$ is naturally a Cauchy sequence in X. Because X is complete, this sequence converges to certain $y_0 \in X$. Because Y is closed, therefore we have $y_0 \in Y$. Because the norms are the same, $\lim_{n\to\infty} y_n = y_0$ in X means exactly $\lim_{n\to\infty} y_n = y_0$ in Y. This proves that a closed subspace of a Banach space is complete.

We further note that if Y is closed for vector addition and multiplication by scalars, i.e., $\forall y_1, y_2 \in Y, \lambda \in \mathbb{R}$, it holds $y_1 + y_2 \in Y, \lambda y_1 \in Y$, then Y is a vector space, and a Banach space. In extending the Banach fixed point theorem, however, the sub-vectorspace is not necessary.

Checking the five steps in the proof of the Banach fixed point theorem, we may obtain the unique fixed point in Y through the iteration procedure.

1.2.2 Applications to ODE system and linear algebraic system

Now we apply this theorem to an ordinary differential equation

$$x'(t) = f(t, x), \quad x(t_0) = x_0. \tag{1.29}$$

Here we assume that $f(t, x)$ is bounded and Lipschitz with respect to x, namely, there are $c, k > 0$, such that

$$|f(t, x)| \leqslant c, \quad |f(t, x) - f(t, y)| \leqslant k |x - y|. \tag{1.30}$$

Taking a time interval $J = [t_0 - \beta, t_0 + \beta]$, we consider the Banach space $C(J)$. We further restrict to a subspace, in the geometrical sense,

$$\tilde{C}(J) = \{x(t) \in C(J) | \, |x(t) - x_0| \leqslant c\beta\}. \tag{1.31}$$

It is straightforward to show that $\tilde{C}(J)$ is a closed subspace. Now we construct an operator $T : \tilde{C}(J) \to \tilde{C}(J)$ by

$$Tx(t) = x_0 + \int_{t_0}^{t} f(\tau, x(\tau)) \mathrm{d}\tau. \tag{1.32}$$

First, we verify $Tx \in \tilde{C}(J)$. In fact, we have

$$|Tx(t_1) - Tx(t_2)| = \left| \int_{t_1}^{t_2} f(\tau, x(\tau)) \mathrm{d}\tau \right| \leqslant c|t_1 - t_2|. \tag{1.33}$$

In particular, we have

$$|Tx(t) - x_0| = \left| \int_{t_0}^{t} f(\tau, x(\tau)) \mathrm{d}\tau \right| \leqslant c |t - t_0| \leqslant c\beta. \tag{1.34}$$

Secondly, we check if it is a contraction. Let $x, y \in \tilde{C}(J)$, we have

$$|Tx(t) - Ty(t)| = \left| \int_{t_0}^{t} [f(\tau, x(\tau)) - f(\tau, y(\tau))] \mathrm{d}\tau \right| \leqslant k\beta \|x - y\|. \tag{1.35}$$

If we take $\beta < 1/k$, then T is a contraction.

Now by the Banach fixed point theorem, there is a unique fixed point $x(t) \in \tilde{C}(J)$ for T, namely,

$$x(t) = x_0 + \int_{t_0}^{t} f(\tau, x(\tau)) \mathrm{d}\tau. \tag{1.36}$$

This gives a solution to the original ODE. We notice that the meaning of a solution needs generalization at a time t^* if $f(t, x(t))$ is discontinuous with respect to t at t^*.

The fixed point theorem also provides a way to solve the ODE by iteration. In fact, taking any $x_0(t) \in \tilde{C}(J)$, we may iterate with

$$x_{n+1}(t) = x_0 + \int_{t_0}^{t} f(\tau, x_n(\tau)) \mathrm{d}\tau. \tag{1.37}$$

The limit gives the fixed point, and therefore the solution. This is called the Picard iteration.

This result may be directly extended to the case of an ODE system.

The fixed point theorem has more general applications. For instance, we may adopt it to solve linear algebraic equations. Consider the space $\mathbb{R}^n$, with a norm

$$\|x\| = \max_{1 \leqslant j \leqslant n} |x_j|. \tag{1.38}$$

A linear mapping T takes the form of $Tx = Ax + b$, where $A = (a_{ij})$, and $b = (b_j)$.

$$\|Tx - Ty\| = \|A(x - y)\| = \max_{1 \leqslant j \leqslant n} \left| \sum_{k=1}^{n} a_{jk}(x_k - y_k) \right| \leqslant \|x - y\| \max_{1 \leqslant j \leqslant n} \sum_{k=1}^{n} |a_{jk}|. \tag{1.39}$$

Therefore, if a row sum criterion $\max_{1 \leqslant j \leqslant n} \sum_{k=1}^{n} |a_{jk}| < 1$ is satisfied, then T is a contraction, and a unique fixed point may be obtained through iteration. The fixed point solves a linear equation $(I - A)x = b$.

Chapter 1 Qualitative Theory for ODE Systems

Consider a linear algebraic system $Cx = d$. We reformulate it with $C = E - F$. Thus we have $Ex = Fx + d$, or $x = E^{-1}(Fx + d)$. So, we shall take $A = E^{-1}F$, and $b = E^{-1}d$.

The key issue then becomes the decomposition of C. To form a good iteration, there are three aspects one needs to consider. First, E should be easy to invert. Secondly, $E^{-1}F$ should be a contraction. Thirdly, a small contraction constant α is preferred.

Let us describe the Jacobian iteration by taking $E = \text{diag}(c_{ii})$, which is the diagonal part of C. To use this method, we require that the diagonal entries dominate corresponding rows, namely,

$$\sum_{k=1, k \neq j}^{n} |c_{jk}| < |c_{jj}|, \quad \forall j = 1, \cdots, n. \tag{1.40}$$

This immediately leads to $\max_{1 \leq j \leq n} \sum_{k=1}^{n} |a_{jk}| < 1$, and the operator T is a contraction. The iteration method actually can be written explicitly, since E is easily inverted,

$$x_j^{(m+1)} = \frac{1}{c_{jj}} \left(d_j - \sum_{k=1, k \neq j}^{n} c_{jk} x_k^{(m)} \right). \tag{1.41}$$

As an example, we find the formulation of the Jacobian iteration for the following matrix,

$$C = \begin{bmatrix} 10 & 5 & 4 \\ 7 & 10 & 2 \\ 1 & 3 & 10 \end{bmatrix}. \tag{1.42}$$

It is a diagonally dominated matrix. We take for the Jacobian iteration,

$$E = \begin{bmatrix} 10 & 0 & 0 \\ 0 & 10 & 0 \\ 0 & 0 & 10 \end{bmatrix}, \quad F = \begin{bmatrix} 0 & -5 & -4 \\ -7 & 0 & -2 \\ -1 & -3 & 0 \end{bmatrix}. \tag{1.43}$$

Then the contractive matrix is

$$A = E^{-1}F = \begin{bmatrix} 0 & -\frac{1}{2} & -\frac{2}{5} \\ -\frac{7}{10} & 0 & -\frac{1}{5} \\ -\frac{1}{10} & -\frac{3}{10} & 0 \end{bmatrix}. \tag{1.44}$$

We remark that there are other ways to decompose the matrix C and form an iteration. For instance, we may take the upper triangular sub-matrix as E, and the lower triangular matrix as F, respectively,

$$E = \begin{bmatrix} 10 & 5 & 4 \\ 0 & 10 & 2 \\ 0 & 0 & 10 \end{bmatrix}, \quad F = \begin{bmatrix} 0 & 0 & 0 \\ -7 & 0 & 0 \\ -1 & -3 & 0 \end{bmatrix}. \tag{1.45}$$

1.3 Critical point

As a trajectory goes locally tangent to the vector field around a regular point, the critical points play an important role in the whole picture of the solutions to an autonomous ODE system. Understanding of critical points helps us capturing the key features of the trajectories.

Around a critical point x_0, we perform Taylor expansion as follows,

$$f(x) = \nabla f(x_0) \cdot (x - x_0) + o(|x - x_0|). \tag{1.46}$$

The Jacobian matrix $A = \nabla f(x_0)$, which is assumed to be invertible, determines the topological structure of trajectories near the critical point. From linear algebra, we know that A is similar to its Jordan normal form. That is, there is an invertible matrix P, such that $PAP^{-1} = B$, where B consists of diagonal entries, as well as Jordan blocks. Now let $y = P(x - x_0)$, the governing equation for y is

$$y' = By + o(|y|). \tag{1.47}$$

While the invertible matrix P may correspond to a combination of rotation, reflection and dilation, the trajectories of the y-system are topologically the same as the x-system. Moreover, it may be proved that except for certain degenerate cases, the topological structure of the trajectories is the same as the linearized system

$$f(x) = \nabla f(x_0) \cdot (x - x_0), \text{ or } y' = By. \tag{1.48}$$

For the sake of clarity, we confine ourselves to the case of a second order ODE system, that is, A and B are 2 by 2 matrices. For the linearized problem (1.48), there are following possibilities. See Fig. 1.4.

- If $B = \begin{pmatrix} \lambda & \\ & \mu \end{pmatrix}$ with $\lambda > \mu > 0$, then the critical point is a source (unstable node), where all trajectories leave x_0. The case with $\mu > \lambda > 0$ is also a source, with a interchanged role of y_1 and y_2.

The general solution is $y = \begin{pmatrix} c_1 e^{\lambda t} \\ c_2 e^{\mu t} \end{pmatrix}$, where c_1, c_2 are arbitrary constants. For $c_1 = 0$, the trajectory is a straight line $y_1 = 0$. For $c_1 \neq 0$, a solution may be expressed in the phase plane as $y_2 = c y_1^{\mu/\lambda}$. In a degenerate or special case $\mu = \lambda$, the trajectories are straight lines. Otherwise, they are curves. A representative case is $\mu = 2\lambda$, for which we have parabolas. Finally, we need to check the direction in a trajectory. Noticing that the absolute values of y_1, y_2 increase as time evolves, each trajectory points outward from the origin.

After we obtain a trajectory in terms of y, we may draw it in the x plane by rotation and distortion, according to the inverse transform of $y = P(x-x_0)$.

- If $B = \begin{pmatrix} \lambda & \\ & \mu \end{pmatrix}$ with $\lambda < \mu < 0$, then the critical point is a sink (stable node), where all trajectories go towards x_0.

The solution takes the same form as the previous case. Moreover, we may consider the time-reversed system with $\tilde{t} = -t$. The critical point then becomes a source. Hence a trajectory remains the same, but with opposite direction.

We remark that the case with $\mu < \lambda < 0$ is similar.

- If $B = \begin{pmatrix} \lambda & \\ & \mu \end{pmatrix}$ with $\lambda > 0 > \mu$, then the critical point is a saddle, where all trajectories go towards and then leave away from x_0. The solution takes the previous exponential form. We remark that the case with $\mu > 0 > \lambda$ is similar.

- If $B = \begin{pmatrix} a & b \\ -b & a \end{pmatrix}$ with $a > 0$, then the critical point is an unstable focus, where all trajectories wind away from x_0. The solution is $y = \begin{pmatrix} c_1 e^{at} \cos bt + c_2 e^{at} \sin bt \\ c_2 e^{at} \cos bt - c_1 e^{at} \sin bt \end{pmatrix}$ for certain coefficients c_1 and c_2.

- If $B = \begin{pmatrix} a & b \\ -b & a \end{pmatrix}$ with $a < 0$, then the critical point is a stable focus, where all trajectories wind towards x_0. The solution takes the same form as that for the unstable focus.

- If $B = \begin{pmatrix} a & b \\ -b & a \end{pmatrix}$ with $a = 0$, then the critical point is a center, where all trajectories wind around x_0. The solution is $y = \begin{pmatrix} c \sin b(t - t_0) \\ c \cos b(t - t_0) \end{pmatrix}$.

When we say toward/away from, we refer to the direction of the trajectory (particle) when time t increases. In the figures, we plot arrows along the direction of increasing t. It is also clear from

$$\frac{dx_i}{dt} = f_i(x) \tag{1.49}$$

that it takes infinite time to enter an unstable critical point, or leave a stable one.

The above classification is not exclusive. There are degenerate cases. For instance, we consider the following situation,

$$B = \begin{bmatrix} a & 1 \\ 0 & a \end{bmatrix}. \tag{1.50}$$

It may be shown that the critical point is a stable node if $a < 0$, and an unstable node if $a > 0$. The trajectories look qualitatively like those for the sources and sinks.

For a nonlinear system, it is not easy to obtain a global picture. However, it may be shown that around a critical point except for the center, the structure keeps the same as the linearized system. We also mention that the degenerate cases, such

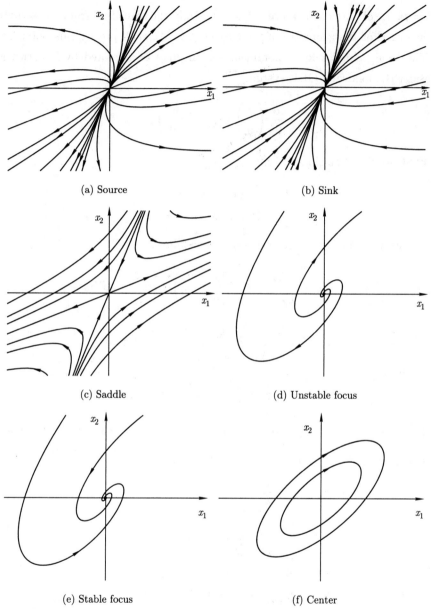

Figure 1.4 Critical points for second order ODE systems.

as $\lambda = \mu$, are of particular interest when we study the transition from one structure to the other for the geometry of the trajectories. We shall precise this point later.

The local behavior for a nonlinear system may be explored by linearization. Consider the following ODE system,

$$\frac{d}{dt}\begin{pmatrix} x \\ y \end{pmatrix} = \begin{pmatrix} f(x,y) \\ g(x,y) \end{pmatrix}. \tag{1.51}$$

First, we find (x_0, y_0) satisfying

$$\begin{cases} f(x_0, y_0) = 0, \\ g(x_0, y_0) = 0. \end{cases} \tag{1.52}$$

Secondly, we perform Taylor expansion as follows,

$$\begin{pmatrix} f(x,y) \\ g(x,y) \end{pmatrix} = \nabla_{(x,y)} \begin{pmatrix} f \\ g \end{pmatrix}(x_0, y_0) \begin{pmatrix} x - x_0 \\ y - y_0 \end{pmatrix} + o(|x - x_0|, |y - y_0|). \tag{1.53}$$

Thirdly, we let $A = \nabla_{(x,y)} \begin{pmatrix} f \\ g \end{pmatrix}(x_0, y_0) \begin{pmatrix} x - x_0 \\ y - y_0 \end{pmatrix}$ and $\begin{cases} \tilde{x} = x - x_0 \\ \tilde{y} = y - y_0 \end{cases}$. Then we obtain

$$\frac{d}{dt}\begin{pmatrix} \tilde{x} \\ \tilde{y} \end{pmatrix} = A \begin{pmatrix} \tilde{x} \\ \tilde{y} \end{pmatrix}. \tag{1.54}$$

Fourthly, we find an invertible matrix P, such that $PAP^{-1} = B$, where B consists of diagonal entries, as well as Jordan blocks. Then we let $\begin{pmatrix} u \\ v \end{pmatrix} = P \begin{pmatrix} \tilde{x} \\ \tilde{y} \end{pmatrix}$. This gives a canonical form

$$\frac{d}{dt}\begin{pmatrix} u \\ v \end{pmatrix} = B \begin{pmatrix} u \\ v \end{pmatrix}. \tag{1.55}$$

With the canonical form, the structure of the trajectories around a critical point can be analyzed according to the previous classification.

1.4 Plane analysis for the Duffing equation

Trajectories of a second order autonomous system lie in the phase plane $\mathbb{R}^2$.

In a linear system, there is either only one critical point, or a line of critical points. Closed orbits exist around a center, with the same period. Complexity may arise in a nonlinear system. First of all, there can be more than one isolated critical points. Secondly, it is possible that a trajectory connects two critical points, or forms a loop from and toward the same critical point. The former is a heteroclinic orbit, and the latter a homoclinic one.

We consider an autonomous system

$$\begin{cases} x' = f(x,y), \\ y' = g(x,y). \end{cases} \quad (1.56)$$

Eliminating the time variable by dividing the two equations, we easily obtain the equation that governs the trajectories in the phase plane,

$$\frac{dy}{dx} = \frac{g(x,y)}{f(x,y)}. \quad (1.57)$$

In many important cases, this equation can be readily integrated. This may provide substantial understanding to the underlying physical system.

As an example, we perform plane analysis to the following Duffing equation,

$$u - u^3 + u'' = 0. \quad (1.58)$$

This equation arises in many applications. For instance, we consider the Lagrangian that describes a particle motion in a double-humped potential field,

$$L = \frac{\dot{u}^2}{2} - \left(\frac{u^2}{2} - \frac{u^4}{4}\right). \quad (1.59)$$

The dynamics is governed by the Euler-Lagrange equation

$$\frac{d}{dt}\left(\frac{\partial L}{\partial \dot{u}}\right) = \frac{\partial L}{\partial u} = -u + u^3. \quad (1.60)$$

Or, equivalently,

$$u'' = -u + u^3. \quad (1.61)$$

This gives the Duffing equation.

Another derivation of this motion is to express the corresponding total energy as a Hamiltonian, with u the displacement and v the velocity (momentum).

$$H(u,v) = \frac{v^2}{2} + \left(\frac{u^2}{2} - \frac{u^4}{4}\right). \tag{1.62}$$

The Hamilton equation reads

$$\frac{du}{dt} = \frac{\partial H}{\partial v} = v, \quad \frac{dv}{dt} = -\frac{\partial H}{\partial u} = -u + u^3. \tag{1.63}$$

Substituting the first equation into the second one, we obtain the Duffing equation.

In the following we present two ways to explore the Duffing equation.

First, we recast it into the ODE system (1.63). There are three critical points, i.e., $(u_+, 0) = (1, 0), (u_0, 0) = (0, 0)$ and $(u_-, 0) = (-1, 0)$.

At a critical point $(u^*, 0)$, the Jacobian matrix is

$$J(u^*, v^*) = \begin{bmatrix} 0 & 1 \\ -1 + 3(u^*)^2 & 0 \end{bmatrix}. \tag{1.64}$$

An eigenvalue λ satisfies

$$\lambda^2 + (1 - 3(u^*)^2) = 0. \tag{1.65}$$

The eigenvector is found to be $(1, \lambda)^T$.

At $(0, 0)$, the eigenvalues are $\pm i$. Linear theory suggests that it is a center. However, for a nonlinear system, a trajectory at the vicinity of such a critical point may not close in general due to the nonlinear terms. It hence becomes a focus, either stable or unstable. We refer to this change of critical point type due to nonlinearity as the structural instability. For the current system, however, we observe there is a symmetry $(u, v; t) \to (u, -v; -t)$. More precisely, if $(u(t), v(t))$ is a solution, so is $(u(-t), -v(-t))$. Therefore, if there is a segment in the upper half plane $v \geq 0$, there must be a symmetric one in the lower half plane $v \leq 0$. When this segment has two intersection points with the u-axis, a closed orbit is formed. Hence the critical point $(0, 0)$ is actually a center.

Actually, we further notice another symmetry for this system $(u, v; t) \to (-u, v, -t)$. That is, if $(u(t), v(t))$ is a solution, so is $(-u(-t), v(-t))$. Therefore, the trajectories have symmetry not only with respect to the u-axis, but also to the v-axis.

Moreover, from $u' = v$, we know that along a trajectory u increases in the upper half plane, and decreases in the lower half plane. This helps determining the time direction along a trajectory. In particular, as $u' = 0$ on the v-axis, all trajectories are locally perpendicular to it.

Next, at $(1, 0)$, the eigenvalues are $\pm\sqrt{2}$. The eigenvectors are $(1, \sqrt{2})^T$ and $(1, -\sqrt{2})^T$, respectively. This means that the stable and unstable manifolds emanate from the saddle with an angle of $\pm \arctan \sqrt{2}$. To see this, we perform a linearization around this saddle point as follows.

Let $(\tilde{u}, \tilde{v}) = (u - u_+, v - v^*)$. Taylor expansion for the source term around $(1, 0)$ gives

$$\begin{pmatrix} v \\ -u + u^3 \end{pmatrix} = J(u_+, v^*) \begin{pmatrix} \tilde{u} \\ \tilde{v} \end{pmatrix} + o(\|(\tilde{u}, \tilde{v})\|). \tag{1.66}$$

Therefore the linearized equations read

$$\frac{d}{dt} \begin{pmatrix} \tilde{u} \\ \tilde{v} \end{pmatrix} = J(u_+, v^*) \begin{pmatrix} \tilde{u} \\ \tilde{v} \end{pmatrix}. \tag{1.67}$$

Now we take a transform matrix

$$P = \begin{bmatrix} 1 & 1 \\ \sqrt{2} & -\sqrt{2} \end{bmatrix}. \tag{1.68}$$

It holds that

$$JP = P \begin{bmatrix} \sqrt{2} & \\ & -\sqrt{2} \end{bmatrix}. \tag{1.69}$$

Therefore, taking a new set of variables

$$\begin{pmatrix} w_1 \\ w_2 \end{pmatrix} = P^{-1} \begin{pmatrix} \tilde{u} \\ \tilde{v} \end{pmatrix}, \tag{1.70}$$

we obtain a decoupled system

$$\frac{d}{dt} \begin{pmatrix} w_1 \\ w_2 \end{pmatrix} = \begin{bmatrix} \sqrt{2} & \\ & -\sqrt{2} \end{bmatrix} \begin{pmatrix} w_1 \\ w_2 \end{pmatrix}. \tag{1.71}$$

This system is readily solved and leads to

$$\begin{pmatrix} u \\ v \end{pmatrix} \approx \begin{pmatrix} u_+ \\ v^* \end{pmatrix} + C_1 e^{\sqrt{2}t} \begin{pmatrix} 1 \\ \sqrt{2} \end{pmatrix} + C_2 e^{-\sqrt{2}t} \begin{pmatrix} 1 \\ -\sqrt{2} \end{pmatrix}. \tag{1.72}$$

The particular solution with $C_2 = 0$ corresponds to the unstable manifold of $(1, 0)$, which is along the direction of $(1, \sqrt{2})$ in the phase plane. Similarly, the particular solution with $C_1 = 0$ corresponds to the stable manifold of $(1, 0)$, along the direction of $(1, -\sqrt{2})$.

Because of symmetry, similar arguments apply to $(-1, 0)$.

Furthermore, still due to the symmetry, as the segment of unstable manifold from $(-1, 0)$ intersects the v-axis, there is a segment that is symmetric with respect to the v-axis and goes precisely toward the critical point $(1, 0)$ along its stable manifold. Therefore, we obtain a trajectory that starts from $(-1, 0)$ and ends up at $(1, 0)$. This is a heteroclinic orbit, which means the trajectory connects two different critical points.

The symmetry with respect to the u-axis shows that yet another heteroclinic orbit exists in the lower half plane. These two orbits separate the phase plane into two regions. Inside them, all trajectories are closed orbits. No closed orbit exists outside of them. They form a separatrix.

The trajectories are schematically summarized in Fig. 1.5.

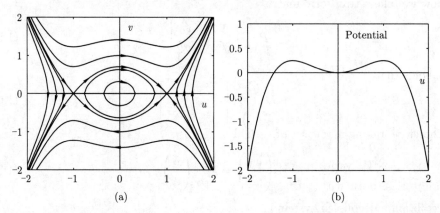

Figure 1.5 The Duffing equation: (a) trajectories in the phase plane; (b) the potential function.

There is an alternative way to analyze the Duffing equation in a more quantitative manner. We multiply the equation by u' and then integrate once to get

$$u^2 - u^4/2 + (u')^2 = C. \tag{1.73}$$

Here $C = 2H$ is an arbitrary constant, representing twice the total energy. We first draw the potential in the right subplot of Fig. 1.5. The maximal potential energy is 1/4, attained at $u = \pm 1$. A local minimal energy $C = 0$ is reached at $u = 0$. Other two displacements that also reach the same total energy are $u = \pm\sqrt{2}$.

For a total energy larger than 1/4, the kinetic energy (hence the absolute value of velocity) increases in the intervals $(-1, 0) \cup (1, +\infty)$, and decreases in $(-\infty, -1) \cup (0, 1)$. Such a whole trajectory, both the positive and negative segments, may reach infinity from both sides.

On the other hand, for a total energy smaller than 1/4, the spatial locations of a trajectory is limited. More precisely, if a particle starts from a position $u_0 \in (-1, 1)$ with zero speed, then it will be confined within the interval between u_0 and $-u_0$. This gives rise to a closed orbit around the center. If the particle starts from a point (with zero velocity) out of this domain, it will go to infinity. Therefore, there are typically three pieces of trajectories in the phase plane for a given total energy between $(0, 1/4)$.

A closed orbit takes $C \in (0, 1/2)$. The amplitude U_0 of the corresponding solution is related to C by $C = U_0^2 - U_0^4/2$. Moreover, the period L is found to be

$$L(U_0) = 4 \int_0^{U_0} [U_0^2 - u^2 - U_0^4/2 + u^4/2]^{-1/2} \, du$$
$$= 4 \int_0^1 [1 - s^2 - U_0^2(1 - s^4)/2]^{-1/2} \, ds. \qquad (1.74)$$

This is an elliptic function. It may be shown that L increases monotonously from 2π to ∞, as U_0 varies from 0 to 1. In particular, when $U_0 = 1$, there are two trajectories, both with infinite period, which means not periodic. They are the heteroclinic orbits.

Here we give a simple Matlab code for solving the Duffing equation.

First, we write a function meta file named duff.m.

```
function F=duff(t,v)
x=v(1);y=v(2);
F=[y;-x+x^3];
```

Then, we run in the command line of Matlab.

```
tspan=[0,20];
v0=[0,1];
[t,v]=ode45('duff',tspan,v0);
x=v(:,1);y=v(:,2);
z=x.^2-x.^4/2+y.^2;
plot3(x,y,z);
```

1.5 Homoclinic orbit and limit cycle

Besides an orbit around a center, there are other types of orbits that form a closed curve in the phase plane, including a homoclinic orbit and a limit cycle.

First, we illustrate a homoclinic orbit by the following example,

$$\begin{cases} x' = y + y(1-x^2)\left[y^2 - x^2(1-x^2/2)\right], \\ y' = x(1-x^2) - y\left[y^2 - x^2(1-x^2/2)\right]. \end{cases} \quad (1.75)$$

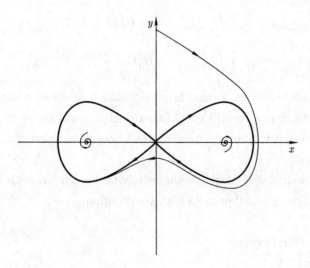

Figure 1.6   Homoclinic orbits and adjacent trajectories.

It may be shown that $(0,0)$ is a saddle; $(1,0)$ and $(-1,0)$ are unstable foci. In fact, at a critical point $(x^*, 0)$, the Jacobian matrix is

$$\begin{bmatrix} 0 & 1 + (1 - (x^*)^2)\left[-(x^*)^2(1 - (x^*)^2/2)\right] \\ 1 - 3(x^*)^2 & (x^*)^2(1 - (x^*)^2/2) \end{bmatrix}. \tag{1.76}$$

Furthermore, it is readily shown that $y^2 = x^2(1 - x^2/2)$ solves the equation. As a result, there are two trajectories. Each of them lies at one side of the y axis, and starts from the saddle and comes back. This is a homoclinic orbit.

Limit cycle is an important type of closed orbits in a nonlinear ODE system. Each trajectory near a limit cycle either goes towards, or leaves from it. Limit cycles may be used to model electric oscillators, or heart beat, among many other applications.

In general, it is extremely hard to detect the existence and the number of limit cycle(s) in a system. In certain cases, the existence may be proved with the help of the following theorem. See Fig. 1.7.

Theorem 1.5.1 (Poincare-Bendixson) *Let B be a ring-shaped domain between two simple closed curves L_1 and L_2. If there is no critical point in this domain, and each trajectory that intersects with one of these two curves enters B, then there exists at least one limit cycle in this domain. The conclusion also holds if the inner boundary curve L_2 shrinks into an unstable node.*

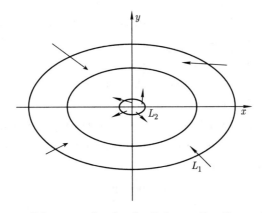

Figure 1.7 Schematic plot for the Poincare-Bendixson theorem.

We remark how to verify that a trajectory enters the domain. As no explicit expression for a solution is known in general, we check the vector field at the boundary curves. If the vector $f(x, y)$ points toward the domain at a non-zero angle, it is possible to show that the trajectory enters the domain. We remark that a limit cycle exists if every trajectory leaves the ring shaped domain.

As an example, we consider the following dynamical system,

$$\begin{cases} x' = -y - x(x^2 + y^2 - 1), \\ y' = x - y(x^2 + y^2 - 1). \end{cases} \tag{1.77}$$

For an outer boundary $L_1 : x^2 + y^2 = 2$, the outer normal at (x_0, y_0) is (x_0, y_0). The source term vector is $(-y_0 - x_0, x_0 - y_0)$, and gives an inner product

$$(x_0, y_0) \cdot (-y_0 - x_0, x_0 - y_0) = -x_0^2 - y_0^2 = -2 < 0. \tag{1.78}$$

Therefore, the source vector points inward on this boundary curve.

Next, we take an inner boundary $L_1 : x^2 + y^2 = 1/2$. It is readily shown that the source term vector $(-y_0 + x_0/2, x_0 + y_0/2)$ points outward, namely, again toward the ring-shaped domain between L_1 and L_2. By the Poincare-Bendixson theorem, the existence of a limit cycle is proved.

Indeed, in the polar coordinates, (1.77) reads

$$\begin{cases} r' = -r(r^2 - 1), \\ \theta' = 1. \end{cases} \tag{1.79}$$

Above two curves are $L_1 : r^2 = 2$, and $L_2 : r^2 = 1/2$. It is easy to check that $r'|_{L_1} = -\sqrt{2} < 0$, whereas $r'|_{L_2} = 1/2\sqrt{2} > 0$. This verifies the assumptions of the Poincare-Bendixson theorem, and a limit cycle exists. In fact, the limit cycle is $r = 1$. All trajectories wind towards this limit cycle, so it is stable.

Next, we consider a more realistic example, namely, the van der Pol equation. It was originally proposed by van der Pol, an electrical engineer working at Philips. In fact, the limit cycle discovered here corresponds to an electronic oscillator to generate a signal (wave) with a special frequency.

For a constant parameter ($\varepsilon > 0$), the equation reads

$$x'' + \varepsilon(x^2 - 1)x' + x = 0. \tag{1.80}$$

It may be recast into an ODE system as follows,

$$\begin{cases} x' = y, \\ y' = -x + \varepsilon(1-x^2)y. \end{cases} \qquad (1.81)$$

The only critical point $(0,0)$ is an unstable node. To show the existence of a limit cycle, it suffices to find an outer boundary curve, along which all the trajectories enter the interior domain.

First, we construct an outer boundary in the following way. We select a point $D(x_D, y_D)$ on the left branch of the curve $-x + \varepsilon(1-x^2)y = 0$. Then we draw the trajectory across this point of the system

$$\begin{cases} x' = y, \\ y' = -x + \varepsilon y. \end{cases} \qquad (1.82)$$

When we take x_D close enough to -1, this trajectory does not intersect with the aforementioned curve on the right of D. It intersects with the y-axis at a point $E(0, y_E)$.

On the other hand, starting from a point $A(0, -y_A)$ on the lower y-axis with $y_A < -y_E$, we may draw a trajectory of the system

$$\begin{cases} x' = y, \\ y' = \varepsilon(1-x^2)y. \end{cases} \qquad (1.83)$$

It intersects with the line $x = -1$ at a point $B(-1, -y_B)$ with $y_B = y_A + 2\varepsilon/3$.

From B, we draw a circle centered at the origin, with a radius $\sqrt{1+y_B^2}$. The circle intersects with the left branch of the curve $-x + \varepsilon(1-x^2)y = 0$ at a point $C(x_C, y_C)$.

We remark that if we take D close enough to the line $x = -1$, and take A far away from $(0, -x_E)$, we may ensure that C lies on the left of D. Meanwhile, the system pertains a symmetry. Therefore, we draw reflected curves with respect to the y-axis, and only need to verify the conditions on the left half of the whole boundary. See Fig. 1.8.

Now we verify the conditions in the Poincare-Bendixson theorem. First, at any point on the segment CD, each trajectory is horizontal because $y' = 0$. Moreover, it points to the right as $x' = y > 0$, therefore enters the domain.

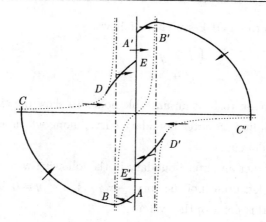

Figure 1.8 Existence of a limit cycle in the van der Pol equation.

Secondly, for the segment DE, again we have $x' = y > 0$. Furthermore, it holds that

$$\left.\frac{dy}{dx}\right|_{DE} = \frac{-x + \varepsilon y}{y} > \left.\frac{dy}{dx}\right|_{vdP} = \frac{-x + \varepsilon(1 - x^2)y}{y}. \tag{1.84}$$

As DE is a monotone curve going from lower-left to upper-right, each trajectory enters the domain.

Thirdly, the segment AB is expressed explicitly by $y = \varepsilon(x - x^3/3) - y_A$. So we have $x' = y < 0$, and

$$\left.\frac{dy}{dx}\right|_{AB} = \frac{\varepsilon(1 - x^2)y}{y} < \left.\frac{dy}{dx}\right|_{vdP} = \frac{-x + \varepsilon(1 - x^2)y}{y}. \tag{1.85}$$

So, again each trajectory enters the domain. We note that a special discussion is needed for the point A, which we omit here.

Finally, for the segment BC, we may construct a function, which we shall discuss in more detail in the next section, $V(x, y) = x^2 + y^2$. Along each trajectory, it is found that

$$\frac{dV}{dt} = 2xx' + 2yy' = 2\varepsilon(1 - x^2)y^2. \tag{1.86}$$

As $x^2 > 1$ on BC, a trajectory takes the direction of decreasing V. Thus it enters the domain.

This ends the proof.

Actually, at the corner point B, the trajectory is horizontal. It is not obvious that the trajectory enters the domain. One may rectify this by taking the AB segment slightly longer beyond the line $x = -1$.

We remark that a limit cycle exists only for a nonlinear system. The Poincare-Bendixson theorem only works for second order ODE's. In higher space dimensions, the geometry can be much more involved. It is usually difficult to show the existence of a limit cycle.

We also remark that a closed orbit may take the form of a limit cycle, an orbit around a center, or a homoclinic orbit. In the mean time, an open orbit may be a heteroclinic orbit, or an orbit that lies between two critical points or infinity.

1.6 Stability and Lyapunov function

Besides critical points and closed orbits, stability is one of the key issues in the qualitative theory. Interests on stability originally emerged from celestial mechanics. One naturally asks: is the galaxy or the solar system stable?

The foundation of stability theory was laid essentially by two classical papers, i.e., *Sur les courbes definies par une equation differentielle* by Poincare, and the dissertation by Lyapunov.

We consider an ODE system

$$x' = f(x), \quad x \in \mathbb{R}^n. \tag{1.87}$$

Stability refers to that of a limit set. A limit set is the set of limit points of a trajectory. More specifically, we consider a trajectory $x = \varphi(t; t_0, x_0)$ that issues from a point (t_0, x_0). A point p is called an ω-limit (or α-limit) point of this trajectory, if there exists a sequence $t_n \to +\infty$ (or $t_n \to -\infty$), such that $\lim_{n \to \infty} |\varphi(t_n; t_0, x_0) - p| = 0$. All such limit points form the limit set of this trajectory. Typically for a second order ODE system, the limit set includes critical point(s), limit cycle, heteroclinic orbit, homoclinic orbit, or a closed orbit around a center. In this section, we shall confine ourselves to the stability of a critical point.

To motivate our analysis, we consider two different situations shown in Fig. 1.9. In both cases, the ball is assumed to stay still if initially so. However, the first one is unstable, in the sense that any small perturbation drives it away from the current equilibrium; whereas the second one is stable, in the sense that the ball remains close to the equilibrium under small perturbation.

In a friction free system with a potential function $V(x)$, the corresponding dynamical system is

$$\begin{cases} \dot{x} = v, \\ \dot{v} = -V'(x). \end{cases} \quad (1.88)$$

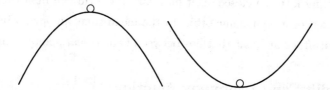

Figure 1.9 Stability of a ball: unstable (left) and stable (right).

We define the total energy as $E(t) = \dfrac{\dot{x}^2}{2} + V(x)$. It may be shown that

$$\frac{dE}{dt} = \dot{x}\ddot{x} + V'(x)\dot{x} = \dot{x}(\dot{v} + V'(x)) = 0. \quad (1.89)$$

The energy conserves. As $V(x)$ attains its minimum at the equilibrium, the particle will remain close to the equilibrium if initially so in the second case of Fig. 1.9. In contrast, $V(x)$ is concave in the first case, and the velocity increases indefinitely after an arbitrarily small perturbation to the equilibrium.

Next, we include friction. The governing equations become

$$\begin{cases} \dot{x} = v, \\ \dot{v} = -V'(x) - \alpha v. \end{cases} \quad (1.90)$$

Correspondingly, we have

$$\frac{dE}{dt} = -\alpha v^2. \quad (1.91)$$

If the potential is convex, the energy decreases, and the particle moves closer and closer to the equilibrium as time elapses.

There are different stabilities defined historically. We list some of these definitions. For simplicity, we consider a critical point $(0,0)$ of a second order ODE system.

- A critical point (0,0) is stable, if $\forall \varepsilon > 0, \exists \delta > 0$, such that $||\varphi(t; t_0, x_0)|| < \varepsilon, \forall t > t_0, \forall ||x_0|| < \delta$.
- A critical point (0,0) is asymptotically stable, if it is stable and $\exists \delta > 0$, such that $\lim_{t \to +\infty} \varphi(t; t_0, x_0) = 0, \forall ||x_0|| < \delta$.
- A critical point (0,0) is exponentially stable, if $\exists \alpha > 0$, for $\forall \varepsilon > 0, \exists \delta > 0$, such that $||\varphi(t, t_0, x_0)|| \leqslant \varepsilon e^{-\alpha(t-t_0)}, \forall ||x_0|| < \delta$.

Indeed, what we have here are uniformly stable, uniformly asymptotically stable, and uniformly exponentially stable, where uniform means the choice of δ is independent of the time t_0.

Stability of a critical point has already been discussed before in a linear system. With the exact solutions, we know that source, saddle and unstable focus are unstable, whereas sink, stable focus and center are stable. However, a critical point of a nonlinear system, which is a center for its linearized system, is usually a focus. Therefore, center is structurally unstable.

To determine the stability of a critical point for a nonlinear system, there are the Lyapunov first method and the Lyapunov second method.

In the first method, one first examines the linear stability of the linearized problem

$$\dot{y} = \nabla f(x_0) \cdot y. \tag{1.92}$$

Linear stability holds if $\nabla f(x_0)$ is negative-definite. Furthermore, the following theorem ensures the nonlinear stability.

Theorem 1.6.1 *If all eigenvalues of $\nabla f(x_0)$ have negative real part, then the critical point x_0 is asymptotically stable for the nonlinear ODE system.*

For the second Lyapunov method, or sometimes called the direct Lyapunov method, we construct a Lyapunov function. Without loss of generality, we assume that the critical point is $x_0 = 0$.

Theorem 1.6.2 *If there exists a Lyapunov function $V(x)$, which is continuously differentiable with respect to x, positive definite, and non-increasing along each trajectory $x(t)$, i.e.*

$$V(0) = 0; \quad V(x) > 0 \text{ for } x \neq 0;$$

$$\left.\frac{dV}{dt}\right|_{x(t)} = \nabla_x V \cdot \frac{dx}{dt} = \nabla_x V \cdot f(x) \leqslant 0 \text{ in a neighborhood } \check{U}(0,\delta).$$

Then the critical point 0 is uniformly stable. If $\left.\dfrac{dV}{dt}\right|_{x(t)} < 0$ in a neighborhood $\check{U}(0,\delta)$, then it is asymptotically stable. If the inequality holds for $\forall x \in \mathbb{R}^n$, and $V(x)$ is radially unbounded, i.e., $\lim\limits_{|x|\to\infty} V(x) \to +\infty$, then we have global (asymptotic or exponential) stability.

We also remark that a sufficient condition for global exponential stability is that there exist positive constants c_1, c_2, c_3, such that

$$c_1 |x|^2 \leqslant V(x) \leqslant c_2 |x|^2, \quad \text{and} \quad \left.\frac{dV}{dt}\right|_{x(t)} \leqslant -c_3 |x|^2.$$

This is illustrated in Fig. 1.10.

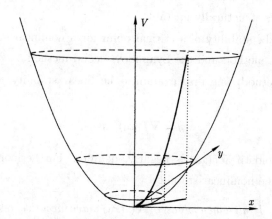

Figure 1.10   Lyapunov function and a trajectory.

A Lyapunov function has two properties. First, it is similar to a norm in the phase space, in a geometrical sense. Secondly, it decreases along with time.

The level curves $V(x) = C$ form a parametrization of the phase plane. These curves, with a decreasing $V(x)$, imply a confinement for the trajectories. Typically, the Lyapunov function is designed in a quadratic form, namely, $V(x) = x^T A x$.

As an example, we consider the motion of a particle with mass $m = 1$ in a conservative force field $\nabla V(x)$. The equation reads, according to Newton's second

law,
$$\ddot{x} = -\nabla V(x). \tag{1.93}$$

Let velocity be $v = \dot{x}$, and rewrite the equation as
$$\begin{cases} \dot{x} = v, \\ \dot{v} = -\nabla V(x). \end{cases} \tag{1.94}$$

It reduces to (1.88) if the space is one dimensional. A critical point must be stationary $v = 0$, and locate at an extremal point of V. Without loss of generality, let us assume the extremal point locates at 0.

Consider the total energy $E = v^2/2 + V(x)$. We construct a function as
$$\mathcal{V}(x, v) = mv^2/2 + V(x) - V(0). \tag{1.95}$$

It turns out that $\dot{\mathcal{V}} = 0$, which means the conservation of energy. So, if we have $V(x) > V(0)$ for a certain neighborhood $\check{U}(0, \delta)$, then $\mathcal{V}$ is a Lyapunov function, and the equilibrium is stable.

This is exactly the Lagrange theorem, which says that an equilibrium position is stable for a conservative force field, if the potential $V(x)$ reaches its local minimum.

1.7 Bifurcation

In applications, a system usually has certain tunable parameters. For instance, there may be a variable resistance or capacitor in an electronic circuit. Consequently, as observed in practice, the system may undergo certain structural change, namely, either its limit set changes or the stability of the limit set changes. Accordingly we observe the change in a physical system when such structural change occurs. As is well known, all that can be observed/measured for a physical system are stable.

To study this in an ODE system, we consider the following parametric system, with $\lambda \in \mathbb{R}$ a parameter,
$$x' = f(x, \lambda). \tag{1.96}$$

At each fixed λ, the analysis in the previous sections applies. More precisely, we first find the critical points. Secondly, we check the type of each critical point using the Jacobian matrix. Thirdly, we draw the representative trajectories schematically.

The first example is the following equation

$$x' = \lambda - x^2. \tag{1.97}$$

There is no critical point for $\lambda < 0$, and two critical points $\pm\sqrt{\lambda}$ appear for $\lambda > 0$. As we have learned, critical points are essential in determining the structure of trajectories. Therefore, we expect a drastic difference when λ crosses 0. In fact, for $\lambda < 0$, because $\lambda - x^2 < 0$, x keeps decreasing towards $-\infty$, regardless to the initial data.

We further remark that the exact solution with initial data $x(t_0) = x_0$ is

$$x(t) = -\mu \tan\left(\mu(t-t_0) - \arctan(x_0/\mu)\right), \quad \mu = \sqrt{-\lambda}. \tag{1.98}$$

It is defined only for a finite time interval. The state variable x tends to $-\infty$ as t approaches $t_0 + (\pi/2 + \arctan(x_0/\mu))/\mu$. We refer to this as a finite time blow-up. For qualitative analysis, we usually ignore this, as the trajectories do not contain much time information except the direction.

On the other hand, for $\lambda = 0$, the exact solution is

$$x = \frac{x_0}{x_0(t-t_0)+1}. \tag{1.99}$$

So, if $x_0 > 0$, x tends to the critical point; and if $x_0 < 0$, x tends to $-\infty$. Different from what we have discussed before, the critical point 0 is neither a source nor a sink. In terms of stability, it is stable from the right semi-axis and unstable from the left semi-axis. It is unstable in a general sense. In fact, if we check the Jacobian $\frac{df}{dx} = -2x^* = 0$. This is a degenerate case.

If λ further increases to $\lambda > 0$, then x tends to $\sqrt{\lambda}$ if $x_0 > -\sqrt{\lambda}$; and tends to $-\infty$ otherwise. The Jacobian $\frac{df}{dx} = -2x^*$ is negative for $x^* = \sqrt{\lambda}$ and positive for $x^* = -\sqrt{\lambda}$. Along with the change in λ, we observe a bifurcation phenomenon, when the controlling parameter λ crosses the bifurcation point $\lambda = 0$.

To illustrate this, we draw a bifurcation diagram in Fig. 1.11. We plot x versus λ for the above first order system. In a multi-dimensional setting, i.e., $x \in \mathbb{R}^n$, we select certain aspect of x for such an illustration, usually an entry x_i, or the norm

$\|x\| = \sqrt{\sum_{i=1}^{n} x_i^2}$. At each representative value for λ, we draw the trajectory with suitable direction of time. For $\lambda > 0$, the stable critical point lies in the solid line (stable branch), whereas the unstable one in the dashed line (unstable branch).

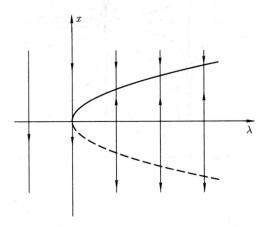

Figure 1.11   Bifurcation diagram: saddle-node bifurcation.

This type of bifurcation is called a tangential bifurcation. It is also called as a saddle-node bifurcation. The meaning becomes clearer if we supplement a decoupled equation to (1.97),

$$\begin{cases} \dot{x} = \lambda - x^2, \\ \dot{y} = -y. \end{cases} \quad (1.100)$$

The second variable converges toward the x-axis exponentially, regardless to the dynamics in x. Now for $\lambda > 0$ we get a saddle $(-\sqrt{\lambda}, 0)$ and a node (sink) $(\sqrt{\lambda}, 0)$. See Fig. 1.12.

Now we consider an example of transcritical bifurcation in the equation

$$x' = \lambda x - x^2. \quad (1.101)$$

We notice that there are always two critical points, $x_1 = 0, x_2 = \lambda$. For $\lambda < 0$, x_1 is stable, and x_2 is unstable. For $\lambda > 0$, x_1 becomes unstable, and x_2 is stable. An exchange of stability occurs. Please refer to the bifurcation diagram in Fig. 1.13.

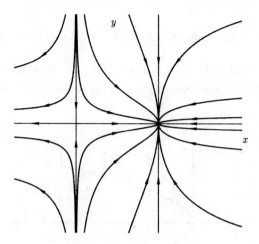

Figure 1.12 Saddle-node bifurcation in a second order system with $\lambda > 0$.

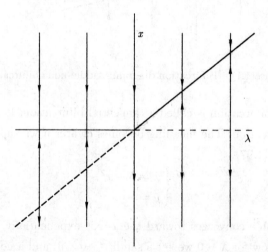

Figure 1.13 Bifurcation diagram: transcritical bifurcation.

Next, consider the supercritical bifurcation in

$$x' = \lambda x - x^3. \tag{1.102}$$

When $\lambda \leqslant 0$, the only critical point is $x_0 = 0$, which is stable. When $\lambda > 0$, there appear two more critical points $x_1 = \sqrt{\lambda}$ and $x_2 = -\sqrt{\lambda}$. These two critical points are both stable, and $x_0 = 0$ becomes unstable. See Fig. 1.14.

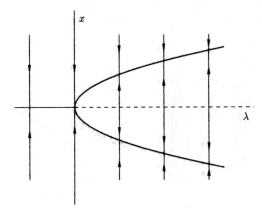

Figure 1.14 Bifurcation diagram: supercritical bifurcation.

The examples we have discussed so far are stationary bifurcations, namely, they concern only with critical points. The structural change may involve closed orbits as well. One most important bifurcation of this type is the Hopf bifurcation. We demonstrate it through a second order ODE system

$$\begin{cases} x' = -y + x(\lambda - x^2 - y^2), \\ y' = x + y(\lambda - x^2 - y^2). \end{cases} \quad (1.103)$$

The only critical point is $(x_0, y_0) = (0, 0)$. The Jacobian matrix is $\begin{pmatrix} \lambda & -1 \\ 1 & \lambda \end{pmatrix}$, which has eigenvalues $\lambda \pm i$. Therefore, it is a stable focus when $\lambda < 0$, and an unstable focus when $\lambda > 0$. In fact, this system may be better understood in polar coordinates $(x, y) = (\rho \cos \theta, \rho \sin \theta)$. It is readily shown that

$$\begin{cases} \rho' = \rho(\lambda - \rho^2), \\ \theta' = 1. \end{cases} \quad (1.104)$$

According to the results of supercritical bifurcation, we know that ρ decreases towards 0 when $\lambda < 0$. Therefore, each trajectory winds counterclockwise towards $\rho = 0$. On the other hand, each trajectory winds counterclockwise towards the periodic orbit $\rho = \sqrt{\lambda}$ for $\lambda > 0$. Thus it is a limit cycle with period 2π.

This bifurcation may be depicted either in a 3-D plot, or a 2-D plot in (ρ, λ)-plane. See Fig. 1.15.

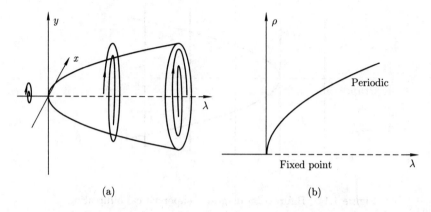

Figure 1.15   Bifurcation diagram: the Hopf bifurcation.

One distinct feature in the Hopf bifurcation lies in the fact that the eigenvalues cross the real axis. Correspondingly a periodic orbit (limit cycle) emerges out of a critical point.

1.8   Chaos: Lorenz equations and logistic map

So far, qualitative theory provides a satisfactory understanding of ordinary differential equations. In particular, plane analysis provides a clean argument for second order autonomous systems. For systems with more degrees of freedom, local behavior is described by the critical point approach. People once believed that the whole picture is essentially similar to that of a second order system.

But this is not true. Topologically speaking, the structure of $\mathbb{R}^3$ is quite different from that of $\mathbb{R}^2$. In the setting of an ODE system, Lorenz put forward an excellent example to illustrate this difference.

Lorenz tried to disprove the mid-term predictability of weather system. Weather system is largely dominated by convection and heat conduction, and may be described by the Rayleigh-Benard equations,

$$\begin{cases} \dfrac{\partial}{\partial t}\left(\nabla^2\psi\right) = -\dfrac{\partial\left(\psi,\nabla^2\psi\right)}{\partial(x,z)} + \nabla^4\psi + \dfrac{R_a}{\sigma}\dfrac{\partial\theta}{\partial x}, \\ \dfrac{\partial}{\partial t}\theta = -\dfrac{\partial\left(\psi,\theta\right)}{\partial(x,z)} + \dfrac{\partial\psi}{\partial x} + \dfrac{1}{\sigma}\nabla^2\theta. \end{cases} \quad (1.105)$$

Here ψ is the stream function, θ is the temperature, R_a is the Rayleigh number, and σ is the Prandtl number. The velocities are recovered from

$$u = -\dfrac{\partial\psi}{\partial z}, \quad w = \dfrac{\partial\psi}{\partial x}. \quad (1.106)$$

Lorenz made a truncation

$$\begin{cases} \psi = \left(\dfrac{\pi^2 + a^2}{\pi a \sigma}\right)\sqrt{2}X(\tau)\sin ax \sin \pi z, \\ \theta = \dfrac{R_c}{\pi R_a}\left[\sqrt{2}Y(\tau)\cos ax \sin \pi z - Z(\tau)\sin 2\pi z\right], \end{cases} \quad (1.107)$$

with $\tau = \dfrac{R_a}{\sigma}\left(\pi^2 + a^2\right)t$, and $R_c = (\pi^2 + a^2)^3/a^2$ is the critical Rayleigh number.

Lorenz obtained the following system which is later named after him,

$$\begin{cases} X' = -\sigma X + \sigma Y, \\ Y' = -XZ + rX - Y, \\ Z' = XY - bZ, \end{cases} \quad (1.108)$$

with $r = R_a/R_c$, $b = 4\pi^2/(\pi^2 + a^2)$. This system possesses a symmetry $\{X, Y, Z\} \to \{-X, -Y, Z\}$.

Defining a Lyapunov function

$$V(t) = rX^2 + \sigma Y^2 + \sigma(Z - 2r)^2, \quad (1.109)$$

we have

$$V' = 2\sigma(-rX^2 - Y^2 - bZ^2 + 2brZ). \quad (1.110)$$

We may take

$$c = \begin{cases} b^2 r^2/(b - \sigma), & \text{if } b \geqslant 2\sigma,\ \sigma \leqslant 1, \\ b^2 r^2 \sigma/(b - 1), & \text{if } b \geqslant 2,\ \sigma \geqslant 1, \\ 4\sigma r^2, & \text{otherwise}. \end{cases} \quad (1.111)$$

Then $V' \leqslant 0$ for $\forall (X, Y, Z)$ satisfying $V \geqslant c$. Therefore, a trajectory in the ellipsoid cannot escape. This defines a trapping zone.

Now consider the critical points. They solve a nonlinear algebraic system

$$\begin{cases} 0 = -\sigma X + \sigma Y, \\ 0 = -XZ + rX - Y, \\ 0 = XY - bZ. \end{cases} \quad (1.112)$$

When $r < 1$, the only critical point is $(0, 0, 0)$. The Jacobian matrix is

$$J = \begin{bmatrix} -\sigma & \sigma & 0 \\ r & -1 & 0 \\ 0 & 0 & -b \end{bmatrix}, \quad (1.113)$$

which has three eigenvalues with non-positive real-part. Therefore, we expect that in this case, all trajectories point towards this node.

In the following, we make analysis under the condition $\sigma - b - 1 > 0$. When $r > 1$, however, one eigenvalue for the Jacobian matrix at $(0, 0, 0)$ becomes positive. It then turns into a saddle. Meanwhile, there appear two new critical points $(\pm\sqrt{b(r-1)}, \pm\sqrt{b(r-1)}, r-1)$. The eigen-equation for both of them is

$$f(\lambda) = \lambda^3 + (\sigma + b + 1)\lambda^2 + (r + \sigma)b\lambda + 2\sigma b(r - 1) = 0. \quad (1.114)$$

We observe that $f(\pm\infty) = \pm\infty$, and $f'(\lambda) = 3\lambda^2 + 2(b + \sigma + 1)\lambda + b(r + \sigma)$. In particular, when $r = 1$, three eigenvalues are $0, -b, -(\sigma + 1)$. Thus when $0 < r - 1 \ll 1$, there are one real eigenvalue with negative sign, and two complex conjugate roots with negative real parts. These two critical points are stable nodes. As r increases, the real eigenvalue becomes smaller and smaller. On the other hand, sum of three roots are fixed as $-(\sigma + 1 + b)$. Therefore, when r becomes big enough, the real eigenvalue becomes smaller than $-(\sigma + 1 + b)$. The two complex eigenvalues take positive real parts. The critical points are then unstable, each with a one dimensional stable manifold and two dimensional unstable manifold. In the unstable manifold, trajectories wind away from the critical points.

More detailed analysis shows that at certain r_0, there appear a pair of heteroclinic orbits, connecting $(0, 0, 0)$ and the unstable critical points. When these heteroclinic orbits break, chaos appears. See Fig. 1.16.

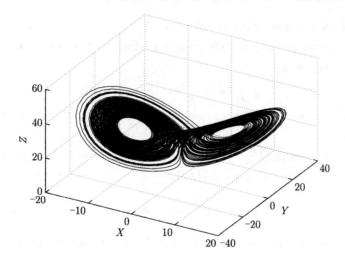

Figure 1.16 A trajectory of the Lorenz equations.

There are many exciting features of chaos. First, there is sensitivity to initial data. All trajectories are bounded within the trapping zone, and there is no aforementioned attractor, such as stable critical point, limit cycle, heteroclinic or homoclinic orbits. Therefore, each trajectory must wind within the ellipsoid forever. This makes a very complex entangled picture. Two trajectories with different yet very close starting points in the phase plane separate from each other after a long run.

Secondly, as pointed out by Lorenz, the deterministic ODE system gives a random-like solution. Because we do not know precisely the initial data, in a long run this imprecision causes different results. In weather forecast, a small perturbation may result in a drastic change after some time, typically in the scale of months.

Another way to study chaos is to go back to the Rayleigh-Benard equation. However, that is too complicated for a substantial understanding. Instead, there are investigations by a better truncation, i.e., with more modes. For instance, by a 14th order ODE system, it was observed that there is a bifurcation process when r increases. More precisely, the system goes to zero equilibrium when $r < 1$; and periodic and quasi-periodic solution appears when r is big enough. When quasi-periodic solution breaks, chaos is observed.

Chaos appears not only in differential equations, but also in mappings. Let us consider the logistic map

$$x_{n+1} = \lambda x_n(1 - x_n). \tag{1.115}$$

Here $\lambda \in [0, 4]$ is a parameter. Assume that $x_0 \in [0, 1]$, then it holds $x_n \in [0, 1]$, $\forall n \in \mathbb{N}$.

When $\lambda = 0$, obviously $x_n = 0$, $\forall n \in \mathbb{N}$. Meanwhile, we notice that the function $f(x) = \lambda x(1-x)$ has a derivative

$$|f'(x)| = |\lambda(1 - 2x)| \leqslant \lambda, \quad \forall x \in [0, 1]. \tag{1.116}$$

Therefore, it is a contraction provided $\lambda < 1$. Accordingly, we know $x_n \to 0$, by virtue of the Banach fixed point theorem.

There is an alternative way to see this. When $\lambda \leqslant \lambda_1 = 1$, $x = \xi_1 = 0$ is the only fixed point of $f(x)$. This fixed point is globally stable. To see this, we notice that for $x \neq 0$, it holds that

$$x_{n+1} \leqslant x_n(\lambda(1-x_n)) < x_n. \tag{1.117}$$

So $\{x_n\}$ forms a decreasing sequence bounded from below by 0. It converges to a certain limit. Due to the continuity of the map, the limit is the fixed point $x = \xi_1 = 0$.

When $\lambda > 1$, $x = 0$ becomes locally unstable, as $f'(0) = \lambda > 1$. There appears another fixed point $\xi_2 = 1 - 1/\lambda$. When $\lambda \in (1, 3)$, from $f'(1 - 1/\lambda) = 2 - \lambda$, we may show local stability for this new fixed point. Actually it is globally stable. To see this, we plot the curve $y = f(x)$ and the line $y = x$ in Fig. 1.17.

When λ exceeds $\lambda_2 = 3$, both fixed points are unstable. However, there appears a 2-periodic state. That is, $f_2(x) = f(f(x))$ has fixed points. They solve

$$x = \lambda(\lambda x(1-x))(1 - \lambda x(1-x)). \tag{1.118}$$

This is a fourth order algebraic equation. Fixed points of $f(x)$ remains to be roots, and there are two new ones,

$$\xi_{3,4} = (1 + \lambda \pm \sqrt{(\lambda + 1)(\lambda - 3)})/2\lambda. \tag{1.119}$$

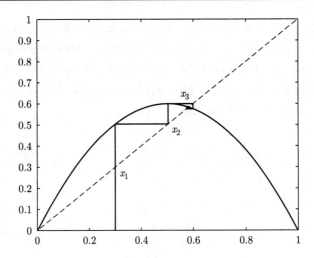

Figure 1.17 Logistic map: $\lambda = 2.4$.

The stability may be studied by analyzing $f_2'(\xi_{3,4})$. By straightforward manipulations, we find that $f_2'(x) = \lambda^2(1-2x)(1-2\lambda x(1-x))$. It turns out that these two roots are stable if $\lambda \leqslant \lambda_3 = 1 + \sqrt{6}$. With an initial value x_0, a typical picture is shown in Fig. 1.18. We observe that when $n \to \infty$, x_n approaches towards this 2-periodic solution $\{\cdots \to \xi_3 \to \xi_4 \to \xi_3 \to \xi_4 \to \cdots\}$.

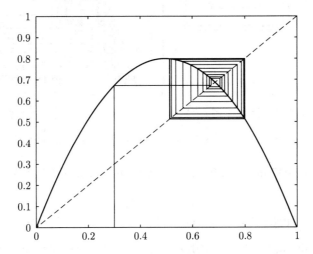

Figure 1.18 Logistic map: 2-periodic solution at $\lambda = 3.2$.

When λ further increases, this 2-periodic solution becomes unstable. It turns out that there appears a stable 4-periodic solution. This corresponds to the rest four fixed points $\xi_{5,6,7,8}$ to $f_4(x) = f_2(f_2(x))$, which is a sixteenth-order polynomial. They are stable if $\lambda \leqslant \lambda_4 = 3.544$. Afterwards, 8-periodic, 16-periodic, 32-periodic, $\cdots$ solutions take place subsequently. See Fig. 1.19.

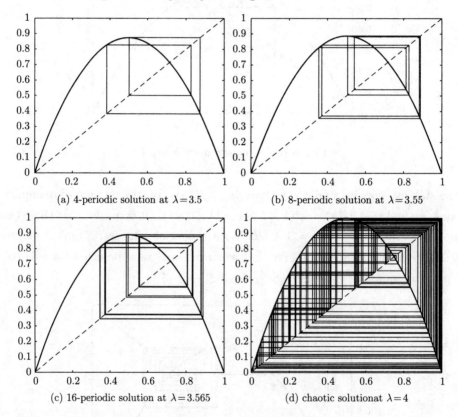

Figure 1.19 Logistic map: period doubling and chaos.

This is a period doubling bifurcation, and the sequence of λ_n takes the following values

$$1, 3, 3.449490, 3.544090, 3.564407, 3.568759, 3.569692, \cdots \qquad (1.120)$$

There is a limit $\lambda_\infty = 3.569945672$. Beyond this threshold, no aforementioned periodic solution is stable, and chaos occurs. There is a uniform constant appears in this bifurcation process

$$\lim_{n\to\infty} \frac{\lambda_n - \lambda_{n-1}}{\lambda_{n+1} - \lambda_n} = 4.669201661. \tag{1.121}$$

It is called the Feigenbaum number, which turns out to be a universal constant for such bifurcations.

Now we consider the special case $\lambda = 4$ in the chaotic regime. Let $x = \sin^2(\pi y/2)$ for $y \in [0,1]$, the iteration reads

$$\sin^2(\pi y_{n+1}/2) = 4\sin^2(\pi y_n/2)\cos^2(\pi y_n/2). \tag{1.122}$$

Noticing the range of y_n, we obtain that

$$y_{n+1} = \psi(y_n) = \begin{cases} 2y_n, & y_n \in [0, 1/2], \\ 2(1-y_n), & y_n \in [1/2, 1]. \end{cases} \tag{1.123}$$

It is plotted in Fig. 1.20.

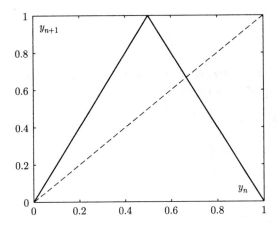

Figure 1.20 Mapping corresponding to the logistic map at $\lambda = 4$.

There are two fixed points, $\eta_1 = 0$ and $\eta_2 = 2/3$, corresponding to $\xi_{1,2}$ respectively. They are also unstable, because the slope is ± 2. In fact, for any $N \in \mathbb{N}$, the N-periodic solutions are unstable.

There is another interpretation of $\psi(y)$. Let $y = 0.p_1 p_2 \cdots$ in binary digits, namely $p_i = 0, 1$ and $y = \sum_i p_i 2^{-i}$, the iteration actually means $\psi(y) = 0.p_2 p_3 \cdots$, or $0.\bar{p}_2 \bar{p}_3 \cdots$. So, this corresponds to a left-shifting.

Sensitivity to initial data is obvious from this interpretation. If $|y - y'| \approx 2^{-k}$, then after k iterations, the difference is amplified to $2^k |y - y'| \approx 1$. Noticing the finite digit representation of a number in computer, we lose the information in initial data completely after certain steps. The second implication is ergodicity, which means that $\{y_n\}$ covers the whole interval of $[0, 1]$. In fact, there are three cases for the exact solution. If $y = \sum_{i=1}^{N} p_i 2^{-i}$, then $y_n = 0$, for $n > N$. If y is a rational number, yet with denominator not in the form of 2^k, then after certain steps, a periodic solution is selected. Finally, if y is irrational, then $\{y_n\}$ covers $(0, 1)$, without repeating. Since there are more irrational numbers than rational ones, we may expect that y_n equally distributed in the interval $[0, 1]$. Correspondingly, x_n distributes in the interval $[0, 1]$ as well, yet not equally.

Assignments

1. Determine if the following systems are autonomous or not. Rewrite them into a system of first order ODE's.

   (a) $ax'' + bx' + cx = 0$;

   (b) $L\dfrac{d^2 I}{dt^2} + R\dfrac{dI}{dt} + \dfrac{1}{C}I = V'(t)$;

   (c) $\begin{cases} x'' = \dfrac{GM}{|x-y|^2}, \\ y'' = -\dfrac{Gm}{|x-y|^2}. \end{cases}$

2. Find the solution to the following systems.

   (a) $L\dfrac{d^2 I}{dt^2} + R\dfrac{dI}{dt} + \dfrac{1}{C}I = 0, \quad I(0) = 0, I'(0) = 1$;

   (b) $\begin{cases} x'' = \dfrac{GM}{|x-y|^2}, & x(0) = R, x'(0) = K, \\ y'' = -\dfrac{Gm}{|x-y|^2}, & y(0) = 0, y'(0) = 0. \end{cases}$

3. Show that $\|x\| = \max_{1 \leq j \leq n} |x_j|$ defines a norm on $\mathbb{R}^n$.

Chapter 1　Qualitative Theory for ODE Systems

4. Is the sequence of functions

$$f_n(t) = \begin{cases} n(t-a), & \text{if } t < a+1/n, \\ 1, & \text{if } t \geqslant a+1/n \end{cases} \tag{1.124}$$

a Cauchy sequence in $C[a,b]$ with the norm $\|f\| = \max\limits_{t\in[a,b]} |f(t)|$?

5. We call a square matrix A as the representation of an operator T under the basis $\{\mathbf{e}_1, \cdots, \mathbf{e}_n\}$, if

$$\begin{aligned} T : \mathbb{R}^n &\to \mathbb{R}^n \\ x &\mapsto y = Ax. \end{aligned} \tag{1.125}$$

Here $x = (x_1, \cdots, x_n)^T = \sum\limits_{i=1}^n x_i \mathbf{e}_i$, $y = (y_1, \cdots, y_n)^T = \sum\limits_{i=1}^n y_i \mathbf{e}_i$. If there is a change of basis $(\mathbf{e}_1, \cdots, \mathbf{e}_n)$ to $(\mathbf{e}'_1, \cdots, \mathbf{e}'_n)$, with $(\mathbf{e}'_1, \cdots, \mathbf{e}'_n) = (\mathbf{e}_1, \cdots, \mathbf{e}_n) Q$, where Q is non-singular, please find the matrix that represents the same operator T.

6. Show that the solution to the ODE given by the fixed point theorem is differentiable, and really solves the ODE, provided that $f(t,x)$ is Lipschitz also with respect to t. Discuss a counter example $f(t,x) = \text{sgn}(t)$.

7. In the proof of local existence for an ODE, we have assumed the boundedness and Lipschitz continuity of $f(t,x)$. However, this is required only locally, for $(x,t) \approx (x_0, t_0)$. Please prove the existence of the solution in $(t_0 - b, t_0 + b)$ by the fixed point theorem under the condition that $\exists a, b, c, k > 0$ with $a \geqslant bc$, $bk \leqslant 1$, such that for $(x,t) \in (x_0 - a, x_0 + a) \times (t_0 - b, t_0 + b)$, it holds that

$$|f(t,x)| \leqslant c, \quad |f(t,x) - f(s,x)| \leqslant k |t-s|, |f(t,x) - f(t,y)| \leqslant k |x-y|. \tag{1.126}$$

8. Find the operator T and matrix E for the Gauss-Seidel iteration

$$x_j^{(m+1)} = \frac{1}{c_{jj}} \left(d_j - \sum_{k=1}^{j-1} c_{jk} x_k^{(m+1)} - \sum_{k=j+1}^n c_{jk} x_k^{(m)} \right). \tag{1.127}$$

Please give a simple sufficient condition for the convergence of the iteration.

9. For a second order system $x' = f(x)$, show that a non-zero vector $f(x) = (f_1(x_1, x_2), f_2(x_1, x_2))^T$ is tangential to the trajectory at the point $x=(x_1, x_2)^T$ in the phase plane.

10. Prove that a trajectory remains unchanged in the phase plane if there undergoes a non-singular time transformation, i.e. $\tau = \tau(t)$ with $\tau'(t) \neq 0$.

11. When we plot the trajectories schematically for the case of a source, we draw all the curves tangential to the y_2-axis, except for the two trajectories along the y_1-axis. Is this correct? Please prove or disprove it.

12. Plot schematically the trajectories for
$$\begin{cases} x' = 3x - y, \\ y' = x + y. \end{cases} \tag{1.128}$$

13. Find the critical points of the system
$$\begin{cases} x' = y(1-x), \\ y' = (1-y)\sin x. \end{cases} \tag{1.129}$$
Determine their types. Can you also plot schematically the trajectories?

14. Show that $(0,0)$ is a saddle; $(1,0)$ and $(-1,0)$ are unstable foci in equation (1.75).

15. Find a curve around the origin to be the inner boundary for the van der Pol equation, from which each trajectory points towards the domain enclosed by this curve and the outer boundary constructed in the lecture.

16. A critical point $(0,0)$ is called attractive, if and $\exists \delta(t_0) > 0$, such that $\lim_{t \to +\infty} \varphi(t; t_0, x_0) = 0$, for $\forall |x_0| < \delta$. Therefore, it is asymptotically stable if it is stable and attractive. Show that being attractive does not imply stability by analyzing the equation
$$\begin{cases} \dfrac{dx}{dt} = y, \\ \dfrac{dy}{dt} = -x(x-1)y - \dfrac{1}{8}x^3. \end{cases} \tag{1.130}$$
(Remark: See Gao W X, Acta Math Sinica, 32(1): 35-41, 1989.)

17. For a linear system $y' = \nabla f(x_0) \cdot y$, if $\nabla f(x_0)$ is negative-definite, find a Lyapunov function.

18. Discuss bifurcation in
$$x' = \lambda + x^2, \tag{1.131}$$
and draw the bifurcation diagram.

Chapter 1 Qualitative Theory for ODE Systems

19. Discuss the subcritical bifurcation in

$$x' = \lambda x + x^3. \tag{1.132}$$

Draw the bifurcation diagram.

20. For the Fitzhugh-Nagumo equations

$$\begin{cases} x' = 3(x + y - x^3/3 + \lambda), \\ y' = -(x - 0.7 + 0.8y)/3, \end{cases} \tag{1.133}$$

compute the critical points, and study their stabilities. Compute when the Hopf bifurcation occurs. Draw a bifurcation diagram in terms of x versus λ.

21. Consider the following equations,

$$\begin{cases} x' = y, \\ y' = x - x^3 + \lambda y + xy/2. \end{cases} \tag{1.134}$$

The critical points are $(0,0), (1,0)$, and $(-1,0)$. Perform numerical computations for $\lambda \in [-0.45, -0.4]$ to investigate when a homoclinic orbit is formed. Plot the representative trajectories for λ around this value. Check the types of the critical points at this value for λ. This is called a homoclinic bifurcation.

22. Please use Matlab to plot some trajectories of the Lorenz equations with $\sigma = 10, r = 28, b = 8/3$.

23. Verify that the Lyapunov function defined by (1.109) satisfies $V' \leqslant 0$ for $V \geqslant c$ in the Lorenz equations.

Chapter 2   Reaction-Diffusion Systems

2.1   Introduction: BVP and IBVP, equilibrium

A partial differential equation (PDE) relates partial derivatives of an unknown variable with itself. The dimension of such a PDE system is n if there are n independent variables ($n \geqslant 2$). The order of a PDE is the order of the highest derivative involved.

In contrast to ODE's, general qualitative theories are not available for PDE's. A property usually relies on the type and even the specific form of an equation. Two highly related tasks are therefore to categorize PDE systems, and to discover their properties.

We start with two-dimensional PDE's of second order. There are basically three types, elliptic, hyperbolic and parabolic.

A linear elliptic PDE that governs a big variety of applications is the following Laplace equation,

$$\Delta u = u_{xx} + u_{yy} = 0. \tag{2.1}$$

In a PDE system that arises from sciences and engineering, there are two types of independent variables, namely, time-like variables and space-like variables. It is not arbitrary to specify a variable as time-like or space-like. For example, in the Laplace equation, we write the independent variables as x and y, implying that they are space-like variables. As a matter of fact, we cannot propose an evolutionary problem for this equation.

To see this, we consider the time evolution of a particular mode with wave number k. That is, we look for a solution in the form of $u(t, x) = f(t)e^{ikx}$ for an evolutionary problem

$$u_{tt} + u_{xx} = 0. \tag{2.2}$$

This gives an equation for $f(t)$,

$$f'' - k^2 f = 0. \tag{2.3}$$

The general solution reads

$$f(t) = C_1 e^{kt} + C_2 e^{-kt}. \tag{2.4}$$

To determine the coefficients C_1 and C_2, we need to prescribe two conditions. For instance, we may assume that

$$f(0) = A, \quad f'(0) = B. \tag{2.5}$$

Or equivalently, we take initial data

$$u(0,x) = A e^{ikx}, \quad u_t(0,x) = B e^{ikx}. \tag{2.6}$$

The solution is then

$$u(t,x) = \big((Ak+B)e^{kt} + (Ak-B)e^{-kt}\big) e^{ikx}/2k. \tag{2.7}$$

Due to linearity, the real part or the imaginary part of this expression also solves the Laplace equation. In particular, we take $A = 1/k^2$ and $B = 1/k$ to get a real-valued solution

$$u(t,x) = e^{kt} \sin kx / k^2. \tag{2.8}$$

Its corresponding initial data read

$$u(0,x) = \sin kx/k^2, \quad u_t(0,x) = \sin kx/k. \tag{2.9}$$

For $k \gg 1$, the initial data is very small, whereas the solution is large at finite time. This gives rise to a blow-up phenomenon. Because a generic initial data contains all wave numbers, immediate blow-up is expected. In another word, even there is a solution which is finite for all time, it is unstable, since any small perturbation (containing short waves) will drive $u(t,x)$ far away from the finite solution. This is Hadamard's example, demonstrating the instability of the Laplace equation as a governing equation for the initial value problem (Cauchy problem).

The Laplace equation suits for boundary value problems. That is, we may solve the equation in a domain (bounded for simplicity) $\Omega \subset \mathbb{R}^2$. The simplest example is $\Omega = [0,a] \times [0,b]$. This describes a plastic membrane fixed by all four sides. To

see this, we recall the separation of variables method. We look for a solution in the form of
$$u(x,y) = f(x)g(y). \tag{2.10}$$
Substituting this into (2.1), we have
$$f''g + fg`` = 0. \tag{2.11}$$
We mention that the derivatives are taken with respect to x and y, respectively. This can be recast into
$$\frac{f''}{f} + \frac{g``}{g} = 0. \tag{2.12}$$
But the first term is a function of x only, and the second one is a function of y only. Therefore, the only possibility is that both are constant. This leads to, for a certain $\lambda \in \mathbb{R}$,
$$f'' = \lambda^2 f, \quad g`` = -\lambda^2 g. \tag{2.13}$$
Or, one may interchange the expression for f and g.

A general solution to (2.1) is
$$u(x,y;\lambda) = (C_1 e^{\lambda x} + C_2 e^{-\lambda x})\cos\lambda y + (C_3 e^{\lambda x} + C_4 e^{-\lambda x})\sin\lambda y. \tag{2.14}$$
Since the equation is linear, superposition holds,
$$u(x,y) = \sum_{\lambda \in \mathbb{R}} u(x,y;\lambda). \tag{2.15}$$
We remark that this is not a proper expression, if there are infinite many λ's in the summation. To determine which λ really appears, we need to specify suitable boundary conditions. In fact, if we put the Dirichlet boundary conditions
$$u(x,0) = u(x,b) = 0, \tag{2.16}$$
then we have
$$u(x,y;\lambda) = (C_3 e^{\lambda x} + C_4 e^{-\lambda x})\sin\lambda y \tag{2.17}$$
with $\sin\lambda b = 0$. Therefore, $\lambda_k = k\pi/b$ with $k \in \mathbb{N}$. So we end up with a general solution
$$u(x,y) = \sum_{k=1}^{\infty} (C_{3,k} e^{\lambda_k x} + C_{4,k} e^{-\lambda_k x})\sin\lambda_k y. \tag{2.18}$$

It is obvious that we need to prescribe further conditions to fix the coefficients. We may impose the Dirichlet boundary conditions, i.e., prescribe the value of $u(0, y)$ and $u(a, y)$; or the Neumann boundary conditions, i.e., prescribe the value of $u_x(0, y)$ and $u_x(a, y)$; or a combination of them.

A more sophisticated way for treating the Laplace equation is the method of Green's function. This can be found in most textbooks on elliptic PDE's.

In case of nonlinear partial differential equations, unfortunately all previous discussions fail. Nevertheless, linear analysis serves as a starting point, and a test-bed for establishing a general theory.

Now we move to reaction-diffusion equations. They describe general applications, not limited to chemical reactions. Systems in biology, physics, mechanics, and even finance and economics may take the form of reaction-diffusion equations.

Similar to the Navier-Stokes equations, these equations may be derived in terms of a balance law.

Let $\rho(\mathbf{x}, t) : \Omega \times \mathbb{R}^+ \to \mathbb{R}$ with $\Omega \in \mathbb{R}^n$ be the particle density function (concentration in chemistry). For any regular subset $B \subseteq \Omega$ as shown in Fig. 2.1, the change of mass $\int_B \rho(\mathbf{x}, t) d\mathbf{x}$ may be induced by two causes:

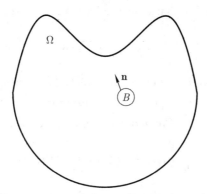

Figure 2.1   Derivation of a balance law.

- The creation of particle within this domain. If we denote creation rate at a point $\mathbf{x} \in \Omega$ and time t as $R(\mathbf{x}, t)$, then this contributes to the change of mass at a rate of $\int_B R(\mathbf{x}, t) d\mathbf{x}$.

- Interchange of mass across the boundary ∂B. This leads to a flux, which means the amount of motion for the particles. More precisely, if the flux density at a position-time pair $(\mathbf{x}, t) \in \Omega \times \mathbb{R}^+$ is $\mathbf{F}(\mathbf{x}, t)$, then at a boundary point $\mathbf{x} \in \partial B$, the amount of incoming particles in a unit time is $-\mathbf{F}(\mathbf{x}, t) \cdot \mathbf{n}$, where $\mathbf{n}$ is the outer normal at this point. Along the whole boundary, the contribution is
$$-\int_{\partial B} \mathbf{F}(\mathbf{x}, t) \cdot \mathbf{n} dA = -\int_B \nabla \cdot \mathbf{F}(\mathbf{x}, t) d\mathbf{x},$$
with help of the divergence theorem.

So, we obtain an equation as follows,

$$\frac{d}{dt}\int_B \rho(\mathbf{x}, t) d\mathbf{x} = -\int_B \nabla \cdot \mathbf{F}(\mathbf{x}, t) d\mathbf{x} + \int_B R(\mathbf{x}, t) d\mathbf{x}. \tag{2.19}$$

Since B is an arbitrary sub-domain, we have a partial differential equation, which is equivalent to the integral equation if everything is smooth (at least continuous) here,

$$\frac{\partial}{\partial t}\rho(\mathbf{x}, t) + \nabla \cdot \mathbf{F}(\mathbf{x}, t) = R(\mathbf{x}, t). \tag{2.20}$$

This is the general form of a balance law. If we change density to another physical quantity, such as energy or momentum, we have similar formulations. A specific application gives specific flux and creation. In particular, according to Fick's law, which is valid in many physical situations, a diffusive process leads to

$$\mathbf{F}(\mathbf{x}, t) = -D\nabla\rho(\mathbf{x}, t). \tag{2.21}$$

Particles tend to go from high density regions to low density regions. The positive coefficient D, or sometimes a positive definite matrix, is the diffusivity, typically measured from experiments. We end up with a reaction-diffusion equation.

$$\frac{\partial}{\partial t}\rho(\mathbf{x}, t) = D\Delta\rho(\mathbf{x}, t) + R(\mathbf{x}, t). \tag{2.22}$$

On the other hand, the creation function depends on specific applications. Phenomenological models are sometimes used. For instance, in an over-simplified chemical reaction model, the concentrations of two reactants A and B are assumed to be $p(x, t)$ and $q(x, t)$, respectively. The reaction is

$$A + B \to 2B + \text{other products}. \tag{2.23}$$

Chapter 2 Reaction-Diffusion Systems

It is natural to assume that the reaction runs according to the richness of the reactants. Therefore, we have

$$\begin{cases} p_t = D_1 p_{xx} - pq, \\ q_t = D_2 q_{xx} + pq. \end{cases} \quad (2.24)$$

Here D_1 and D_2 are the diffusivity of A and B, respectively.

Now we assume that the reaction is taken in a bounded container $x \in [0,1]$. From a physical point view, we need to tell what are the densities at the beginning, and what kind of control we put on the boundary. This is exactly what one needs to impose mathematically. In fact, we specify initial conditions

$$p(x,0) = p_0(x), \quad q(x,0) = q_0(x), \quad (2.25)$$

and boundary conditions

$$\begin{aligned} p(0,t) &= \varphi_0(t), \quad p(1,t) = \varphi_1(t), \\ q(0,t) &= \psi_0(t), \quad q(1,t) = \psi_1(t). \end{aligned} \quad (2.26)$$

There is a mathematical theory why we should propose the problem in this way. Here we present a heuristic argument. Imagine that we solve this PDE system by a numerical method. We discretize the space by a uniform mesh $\{x_i | i = 0, \cdots, N\}$ with mesh size $\Delta x = 1/N$. Taking the solution at a grid point $x_i = i\Delta x$ as $(p_i(t), q_i(t))$, we approximate the derivatives by central differences,

$$\begin{aligned} p_{xx}(i\Delta x, t) &= (p_{i-1}(t) - 2p_i(t) + p_{i+1}(t))/(\Delta x)^2, \\ q_{xx}(i\Delta x, t) &= (q_{i-1}(t) - 2q_i(t) + q_{i+1}(t))/(\Delta x)^2. \end{aligned} \quad (2.27)$$

This gives a set of ODE system as follows,

$$\begin{cases} p_i'(t) = D_1(p_{i+1} - 2p_i + p_{i-1})/(\Delta x)^2 - p_i q_i, \\ q_i'(t) = D_2(q_{i+1} - 2q_i + q_{i-1})/(\Delta x)^2 + p_i q_i, \end{cases} \quad (2.28)$$

for $i = 1, \cdots, N-1$. We cannot allow $i = 0$, or $i = N$, because their equations involve the values at the point $i = -1$ and $i = N+1$, respectively. On the other hand, if we prescribe these boundary data and solve the ODE system, it is necessary and sufficient to provide initial data for each $p_i(0)$ and $q_i(0)$. This justifies, not rigorously, our choice of initial and boundary conditions.

From the above discussions, we notice the importance of boundary and initial data. In fact, for a PDE problem, we should consider the equation and suitable initial/boundary conditions as a whole.

There are several other example systems that we shall explore in this chapter. One is the Fisher equation, which was first derived by Fisher in ecology to describe gene spreading. Here u is the frequency of the mutant gene. It was also developed by Kolmogorov, Petrovskii and Piscounov, so sometimes called the Fisher-KPP equation,

$$u_t = u_{xx} + u(1-u). \qquad (2.29)$$

Another one is the Fitzhugh-Nagumo equations, which govern the conduction of action potentials in certain neural fibers,

$$\begin{cases} u_t = u_{xx} + f(u) - v, \\ v_t = \delta v_{xx} + \sigma u - \gamma v. \end{cases} \qquad (2.30)$$

A PDE system over the domain $\mathbb{R}^n \times \mathbb{R}^+$ with initial condition(s) only is called a Cauchy problem, or an initial value problem (IVP). A PDE system with boundary condition(s) only is called a boundary value problem (BVP). A mixture of them is called an initial boundary value problem (IBVP).

To understand the behavior of such a PDE system, it is much harder than ODE systems. Our knowledge on ODE does help to quite some extent here. In this chapter, we shall illustrate some elementary approaches. We should mention that there is an immense literature of theoretical approaches for PDE's. An exposure to those methods requires a good knowledge of real analysis, functional analysis, and many other mathematical tools.

The simplest thing that we may do is to discard the dependence on the time variable. Then we have steady state equations. For (2.24), they are

$$\begin{aligned} D_1 p_{xx} - pq &= 0, \\ D_2 q_{xx} + pq &= 0. \end{aligned} \qquad (2.31)$$

We may explore this fourth order ODE system by the methods presented in the previous sections. A special case for the limit is, $p = 0$ or $q = 0$. It means the reaction stops when running out of one reactant.

2.2 Dispersion relation, linear and nonlinear stability

For a reaction-diffusion system, we may explore the stability of a simple solution by a dispersion relation approach. That is, we consider a solution $u(x,t) = u_0 + U\exp(\lambda t + i\omega x)$, where u_0 is the equilibrium and $U \ll 1$. Plugging it into the equation, we retain the linear terms only. This typically yields a linear equation. The solvability condition then gives the dependency of λ on ω, which is the dispersion relation. The bigger is the wave number ω, the shorter is the wave length. We remark that physicists use the notation $\exp(i\omega t + ikx)$ instead of $\exp(\lambda t + i\omega x)$, where k is the wave number.

We now consider the Fisher equation (2.29) as an example. The trivial equilibrium state is $u = 0$. Taking $u(x,t) = U\exp(\lambda t + i\omega x)$, we have

$$(\lambda + \omega^2 - 1)Ue^{i\omega x} = -U^2 e^{\lambda t + 2i\omega x}. \tag{2.32}$$

Discarding the nonlinear term on the right hand side, we obtain the sufficient and necessary condition for the existence of a non-zero U as

$$\lambda = 1 - \omega^2. \tag{2.33}$$

This is illustrated in Fig. 2.2. The governing equation for the evolution of this perturbation is therefore a linearized equation, and in our example

$$\tilde{U}_t = \tilde{U}_{xx} + \tilde{U}. \tag{2.34}$$

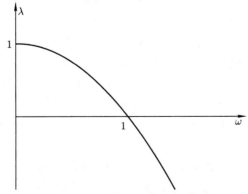

Figure 2.2 Dispersion relation for the Fisher equation.

Since superposition works for a linear equation, we only need to discuss a solution in the form of $\tilde{U} = U \exp(\lambda t + i\omega x)$. Notice that any 'nice' function can be expressed by trigonometric functions, either in terms of Fourier series, or in terms of Fourier expansion.

The dispersion relation indicates the stability, which is crucial in applications. Like in ODE systems, stability of a state means that a small initial perturbation (at $t = 0$) results in small deviation for any time t. The problem is then how we describe a function to be 'small' or 'big'. In an ODE, we discuss a single independent variable function $f(t)$. For an ODE system, it is not this easy. We discuss a vector function $x(t) = (x_1(t), \cdots, x_n(t))$ of a single independent variable t. The size, as used in the stability analysis or the bifurcation study, can be measured by

$$\|x(t)\| = \sqrt{\sum_{i=1}^{n} x_i^2}. \qquad (2.35)$$

For a PDE system, we are talking about the size of a function with two or more independent variables, including one time variable t, and one or more space variables. A commonly used measure is an L^2-norm

$$\|u(x)\|_2 = \sqrt{\int [u(x)]^2 \, \mathrm{d}x}. \qquad (2.36)$$

Here we suppress the dependence on t, and treat it as a parameter.

It is obvious that the definition of stability depends on how we measure the size. An L^2-stability for a ground state, not necessarily to be constant equilibrium $u_0(x,t)$, can be stated as follows. For any $\varepsilon > 0$, there exists $\delta > 0$, such that we have $\|u(x,t) - u_0(x,t)\|_2 < \varepsilon$, provided $\|u(x,0) - u_0(x,0)\|_2 < \delta$. We emphasize that linear stability does not imply nonlinear stability. Nevertheless, in most applications, linear stability is a good sign for nonlinear stability.

With a dispersion relation, we actually perform a linear stability analysis. If there exists an ω, such that the corresponding $Re\lambda > 0$, then the ground state is linearly unstable. In fact, if we perturb the ground state by $U_0 \exp(i\omega x)$ with U_0 very small, we expect an exponential growth at the rate $U \sim U_0 \exp(i\omega x + \lambda t)$, at least for small time t.

A dispersion relation contains more information. If the maximal value of $\text{Re}\lambda$ is reached at a certain ω, then this mode is most probable to be excited. In a real system, this mode is usually the one that is observed. In the meantime, as each mode takes the form of $e^{\lambda t + i\omega x}$, it becomes $e^{i(ct+kx)}$ if $\lambda = ic$ is purely imaginary. The corresponding phase speed is $-c/\omega$, and the group velocity is $dc/d\omega$.

Now we make a simple analysis on (2.24). As we have seen before, the steady states include $(p, q) = (0, Q)$ and $(p, q) = (P, 0)$ with P and Q constant positive concentrations. The linearized equations around $(0, Q)$ read

$$\begin{cases} \tilde{p}_t = D_1 \tilde{p}_{xx} - \tilde{p} Q, \\ \tilde{q}_t = D_2 \tilde{q}_{xx} + \tilde{p} Q. \end{cases} \tag{2.37}$$

Assume that $(\tilde{p}, \tilde{q}) = (\hat{P}, \hat{Q}) \exp(i\omega x + \lambda t)$, then we have

$$\begin{bmatrix} \lambda & 0 \\ 0 & \lambda \end{bmatrix} \begin{bmatrix} \hat{P} \\ \hat{Q} \end{bmatrix} e^{i\omega x + \lambda t} = \begin{bmatrix} -D_1 \omega^2 - Q & 0 \\ Q & -D_2 \omega^2 \end{bmatrix} \begin{bmatrix} \hat{P} \\ \hat{Q} \end{bmatrix} e^{i\omega x + \lambda t}. \tag{2.38}$$

The solvability condition is that the coefficient matrix is singular. That is, we have

$$\lambda_1 = -D_1 \omega^2 - Q < 0; \quad \lambda_2 = -D_2 \omega^2 < 0. \tag{2.39}$$

So, it is stable.

Similarly, for $(P, 0)$, the eigenvalues are

$$\lambda_1 = -D_1 \omega^2 < 0; \quad \lambda_2 = P - D_2 \omega^2. \tag{2.40}$$

Therefore, it is unstable for perturbations with $\omega \leqslant \sqrt{P/D_2}$.

Just as before, nonlinearity may make things completely different. We now illustrate that even the existence of the solution may fail in a nonlinear reaction-diffusion equation.

Consider a spatially periodic problem

$$\begin{cases} u_t = u_{xx} - u + u^2, \\ u(x, 0) = u_0(x) \geqslant 0, \\ u(0, t) = u(\pi, t) = 0. \end{cases} \tag{2.41}$$

It is similar to the Fisher equation but with a different sign for the reaction terms. We assert that the solution will keep positive, as it is initially so. We shall prove this in the next section. Now for such a non-negative solution $u(x,t)$, we associate a function with it,

$$f(t) = \int_0^\pi u(x,t)\sin x\,dx. \tag{2.42}$$

By straightforward calculations, we have

$$\frac{df}{dt} = -2f + \int_0^\pi u^2 \sin x\,dx. \tag{2.43}$$

Noticing that

$$\int_0^\pi u(x,t)\sin x\,dx \leqslant \left(\int_0^\pi u^2 \sin x\,dx\right)^{1/2}\left(\int_0^\pi \sin x\,dx\right)^{1/2}$$

$$= \sqrt{2}\left(\int_0^\pi u^2 \sin x\,dx\right)^{1/2}, \tag{2.44}$$

we have

$$\frac{df}{dt} \geqslant -2f + \frac{f^2}{2}. \tag{2.45}$$

If $f(0) > 4$, then the function $f(t)$ tends to infinity at finite time. This implies that there are something wrong with the function $u(x,t)$. In fact, from

$$[f(t)]^2 \leqslant \int_0^\pi [u(x,t)]^2\,dx \cdot \int_0^\pi (\sin x)^2\,dx \leqslant \frac{\pi}{2}\int_0^\pi [u(x,t)]^2\,dx, \tag{2.46}$$

we find that $\int_0^\pi [u(x,t)]^2\,dx$ must go to infinity at finite time. Therefore, there is a blow-up for u.

2.3 Invariant domain

Though we observed blow-up in a reaction-diffusion system in the previous section, most practical reaction-diffusion systems do admit global and even smooth solutions. One important tool for qualitative study is the comparison principle, and the method of invariant domain/region.

Let us start with a comparison principle for ODE's.

Consider two ODE's

$$x' = f(t,x), \quad x(0) = x_0; \tag{2.47}$$

$$y' = g(t,y), \quad y(0) = y_0. \tag{2.48}$$

If we have $f(t,z) > g(t,z)$ for all $z \in [a,b]$, and initially $x_0 > y_0$, then for all time t such that x and y lie within the interval $[a,b]$, we have $x > y$.

This may be shown by contradiction. If there is a time t^* when we first have $x(t^*) = y(t^*) \in [a,b]$, then at this time, $y'(t^*) \leqslant x'(t^*)$. This contradicts with $f(t^*, x^*) > g(t^*, y^*)$.

Next we consider a reaction-diffusion equation with initial data $u(x,0)$,

$$\begin{cases} u_t = \Delta u + f(u,x,t), & x \in \Omega \subset \mathbb{R}^n, t > 0, \\ \alpha u + \beta \nabla u \cdot \mathbf{n} = g(x,t), x \in \partial\Omega. \end{cases} \tag{2.49}$$

Here $\mathbf{n}$ is the outer normal at the boundary. Throughout this section, we assume that Ω is a 'nice' domain ($\partial\Omega$ is piecewise smooth, and satisfies a cone-condition). We further confine ourselves to either $\alpha = 1, \beta = 0$, namely, $u = g(x,t)$ on $\partial\Omega$; or $\alpha \geqslant 0, \beta = 1$, namely $\alpha u + \nabla u \cdot \mathbf{n} = g(x,t)$ on $\partial\Omega$.

If the following three assumptions hold, we obtain a super-solution $\bar{u}(x,t) \geqslant u(x,t), x \in \Omega, t > 0$,

$$\begin{cases} \bar{u}_t \geqslant \Delta \bar{u} + f(\bar{u},x,t), & x \in \Omega \subset \mathbb{R}^n, t > 0, \\ \alpha \bar{u} + \beta \nabla \bar{u} \cdot \mathbf{n} \geqslant g(x,t), x \in \partial\Omega, \\ \bar{u}(x,0) \geqslant u(x,0). \end{cases} \tag{2.50}$$

This comes from the strong maximum principle for linear parabolic problem, which claims that an extremal point must appear either at the boundary or initially.

In the same fashion, we have a sub-solution $\underline{u}(x,t) \leqslant u(x,t), x \in \Omega, t > 0$, if it holds that

$$\begin{cases} \underline{u}_t \leqslant \Delta \underline{u} + f(\underline{u},x,t), & x \in \Omega \subset \mathbb{R}^n, t > 0, \\ \alpha \underline{u} + \beta \nabla \underline{u} \cdot \mathbf{n} \leqslant g(x,t), x \in \partial\Omega, \\ \underline{u}(x,0) \leqslant u(x,0). \end{cases} \tag{2.51}$$

Let us illustrate how the comparison principles work.

First, with an example of (2.41), it is easy to verify all conditions for $\underline{u}(x,t) = 0$ to be a sub-solution. This proves our previous claim that $u(x,t) \geqslant 0$.

Secondly, we consider an example of the evolutionary Duffing equation,

$$\begin{cases} u_t = u - u^3 + \Delta u, \ x \in \Omega, \\ \nabla u \cdot \mathbf{n} = 0, \quad x \in \partial\Omega. \end{cases} \tag{2.52}$$

If initially $u(x,0) \in [-1,1]$, then we observe that $\bar{u}(x,t) = 1$ is a super-solution, and $\underline{u}(x,t) = 0$ is a sub-solution. That is, we have $u(x,t) \in [-1,1]$.

What happens if the initial data lie between other two constant solutions, e.g., $u(x,0) \in (0,1)$? Then $\bar{u}(x,t) = 1$ is still a super-solution. On the other hand, we have a sub-solution $\underline{u}(x,t) = s(t)$, which solves an ODE $s'(t) = s - s^3$, with initial data $s(0) = \min\limits_{x \in \Omega} u(x,0)$.

The aforementioned comparison principle actually can be interpreted in a geometrical way, which gives rise to the method of invariant domain.

We take a reaction-diffusion system for $\mathbf{u} \in \mathbb{R}^m$,

$$\mathbf{u}_t = D\Delta\mathbf{u} + \mathbf{f}(\mathbf{u}). \tag{2.53}$$

Let $B = \prod\limits_{j=1}^{m} [\underline{u}_j, \bar{u}_j]$ with $\underline{u}_j < \bar{u}_j$, and outer normal $\mathbf{N}$. We emphasize that this is the outer normal in the state space, in contrast to the outer normal $\mathbf{n}$ in the physical space. If it holds that $\mathbf{f} \cdot \mathbf{N} < 0$ for $\mathbf{u} \in \partial B$, then for initial data $\mathbf{u}(x,0) \in B$, and boundary data $\mathbf{u}(x,t)|_{\partial\Omega} \in B$, the solution remains in B. We also allow boundary conditions that guarantee $\nabla u_i \cdot \mathbf{n} < 0$ if $u_i = \bar{u}_i$ and $\nabla u_i \cdot \mathbf{n} > 0$ if $u_i = \underline{u}_i$ for all i. The domain B is called an invariant domain. See Fig. 2.3.

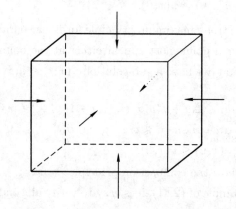

Figure 2.3 Schematic view of an invariant domain.

Now we apply this method to the Fitzhugh-Nagumo equations (2.30), which govern the conduction of action potentials in certain neural fibers.

We take $f(u) = u(1-u)(u-a)$. This allows us to draw an invariant domain in Fig. 2.4. The curves $f(u) = v$ and $\sigma u - \gamma v = 0$ are called nullclines, because they describe the place where the reaction terms vanish.

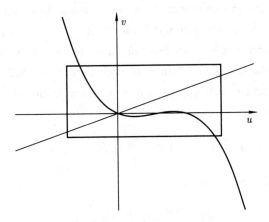

Figure 2.4 An invariant domain for the Fitzhugh-Nagumo equations.

The invariant domain argument applies to a general class of models. For instance, we consider a set of competition equations in population biology,

$$\begin{cases} u_t = d_1 u_{xx} + u(A_1 - B_1 u - C_1 v), \\ v_t = d_2 v_{xx} + v(A_2 - B_2 u - C_2 v). \end{cases} \quad (2.54)$$

Here $u(x,t)$ and $v(x,t)$ denote the population of two species. The constant coefficients $A_i, B_i, C_i, d_i > 0$ for $i = 1, 2$.

For this system, we first use the sub-solution argument to show that $u \geqslant 0, v \geqslant 0$. Then we may find invariant domains according to different situations of the coefficients. For instance, if $A_2 C_1 > A_1 C_2$, $A_2 B_1 > A_1 B_2$, then we may take, with $\varepsilon > 0$ small enough,

$$B = [0, A_1/B_1 + \varepsilon] \times [0, A_2/C_2 + \varepsilon]. \quad (2.55)$$

2.4 Perturbation method

Perturbation around a known state is a general approach that applies to many situations. Such a known state is a ground state, and the analysis is called perturbation analysis. Over the years, mathematicians have developed a fairly general theory for this method. Though nowadays computers are widely used for analyzing nonlinear phenomena, perturbation methods still play an important role toward a better qualitative understanding of nonlinear systems. In certain cases, computer programs can not substitute such a delicate theory.

Taylor expansion of a smooth function bears the essential features of perturbation analysis. For instance, when x is close to $\pi/6$, we take $\varepsilon = x - \pi/6$. A function $\sin x$ is pretty well approximated by $\sin \pi/6$. However, to get a better approximation, we go to a higher order, namely,

$$\sin x = \sin(\pi/6 + \varepsilon) = \sin \pi/6 + \varepsilon A_1 + O(\varepsilon^2). \tag{2.56}$$

Using L'Hospital's rule, we have

$$A = \lim_{\varepsilon \to 0} \frac{\sin x - \sin \pi/6}{\varepsilon} = \cos \pi/6. \tag{2.57}$$

One may go to even higher orders, and end up with the Taylor expansion of $\sin x$ around $x = \pi/6$. We observe that this approach takes a gradually zoom-in procedure, and forms a hierarchy of approximations to the target function $\sin x$.

Now we illustrate the perturbation method for a steady state of a reaction-diffusion system. We consider the Fisher equation (2.29) in the domain $[0, L]$ with boundary conditions

$$u(0, t) = u(L, t) = 0. \tag{2.58}$$

Its steady state solves

$$u_{s,xx} + u_s(1 - u_s) = 0, \quad u_s(0) = u_s(L) = 0. \tag{2.59}$$

Same as before, we multiply the equation by $u_{s,x}$ and integrate by parts to obtain

$$u_{s,x}^2 + u_s^2 - \frac{2}{3}u_s^3 = C. \tag{2.60}$$

This gives, for an orbit passing through $(u_s, u_{s,x}) = (U_0, 0)$,

$$x(u_s) = x_0 \pm \int_0^{u_s} \frac{1}{\sqrt{U_0^2 - u^2 - 2/3(U_0^3 - u^3)}} du. \tag{2.61}$$

This ODE system has two critical points, a center and a saddle. See Fig. 2.5. Noticing that we have a boundary value problem for the reaction-diffusion equation, the only relevant steady state, besides the trivial zero equilibrium $u_s(x) \equiv 0$, is the one that can leave and come back to the $u_{s,x}$ axis exactly within the distance

$$L(U_0) = 2 \int_0^{U_0} \frac{1}{\sqrt{U_0^2 - u^2 - \frac{2}{3}(U_0^3 - u^3)}} du = 2 \int_0^1 \frac{1}{\sqrt{1 - s^2 - \frac{2}{3}U_0(1 - s^3)}} ds. \tag{2.62}$$

It is unique because of the monotonicity of $L(U_0)$. Furthermore, it is observed that

$$L(U_0) \geqslant 2 \int_0^1 \frac{1}{\sqrt{1-s^2}} ds = \pi. \tag{2.63}$$

So we expect that a steady state exists only when L is bigger than π. Hence $L^* = \pi$ is a bifurcation point. We are interested in how the solution looks like for L slightly bigger than π. Using the expression of $x(u_s)$, one certainly knows something about the solution, with the help of the implicit function theorem. However, we seek for better understanding from a perturbation study.

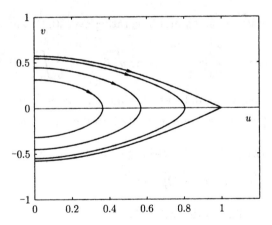

Figure 2.5 Phase plane plot for steady states of the Fisher equation.

To this end, we first let $y = x/L$, and denote u_s as v for ease of presentation. This change of independent variable makes L a controlling parameter in the equation.

The new equation under investigation is

$$v_{yy} + L^2 v(1-v) = 0, \quad v(0) = v(1) = 0. \tag{2.64}$$

Choosing the ground state to be $v^* = 0$, the linearized equation around this solution satisfies

$$v_{yy} + L^2 v = 0, \quad v|_{y=0,1} = 0. \tag{2.65}$$

It has a non-trivial solution only when $L \geqslant L^* = \pi$. This agrees with the previous discussions.

Now we make an ansatz, i.e., claim the form of the solution as

$$L = L^* + \varepsilon, \quad v(y) = v_0(y) + \varepsilon v_1(y) + \varepsilon^2 v_2(y) + \varepsilon^3 v_3(y) + \cdots. \tag{2.66}$$

First, we show that $v_0(y) = 0$. To this end, we substitute the ansatz to the equation. As the $O(\varepsilon)$ terms can not balance with the leading order terms, we have

$$\begin{cases} v_0'' + \pi^2 v_0(1 - v_0) = 0, \\ v_0(0) = v_0(1) = 0. \end{cases} \tag{2.67}$$

Multiplying both sides by $\sin \pi y$ and integrating over $[0,1]$, we obtain

$$0 = \int_0^1 \left(v_0'' \sin \pi y + \pi^2 v_0 - \pi^2 v_0^2 \sin \pi y \right) dy = -\pi^2 \int_0^1 v_0^2 \sin \pi y \, dy. \tag{2.68}$$

Because $\sin \pi y$ is positive in $[0,1]$, we conclude that $v_0(y) = 0$.

Next, substituting the rest of the ansatz into the equation, we have

$$\left[\varepsilon v_1''(y) + \varepsilon^2 v_2''(y) + \varepsilon^3 v_3''(y) \right] + (\pi^2 + 2\pi\varepsilon + \varepsilon^2) \left[\varepsilon v_1(y) + \varepsilon^2 v_2(y) + \varepsilon^3 v_3(y) \right]$$
$$\cdot \left[1 - \varepsilon v_1(y) - \varepsilon^2 v_2(y) - \varepsilon^3 v_3(y) \right] + o(\varepsilon^3) = 0. \tag{2.69}$$

We rearrange it according to the order of ε,

$$\varepsilon \left[v_1''(y) + \pi^2 v_1(y) \right]$$
$$+ \varepsilon^2 \left[v_2''(y) + \pi^2 v_2(y) + 2\pi v_1(y) - \pi^2 v_1^2(y) \right]$$
$$+ \varepsilon^3 \left[v_3''(y) + \pi^2 v_3(y) + 2\pi v_2(y) + v_1(y) - 2\pi^2 v_1(y) v_2(y) - 2\pi v_1^2 \right] + o(\varepsilon^3) = 0. \tag{2.70}$$

Chapter 2   Reaction-Diffusion Systems

Same as before, we have

$$v_1''(y) + \pi^2 v_1(y) = 0,$$
$$v_2''(y) + \pi^2 v_2(y) + 2\pi v_1(y) - \pi^2 v_1^2(y) = 0, \qquad (2.71)$$
$$v_3''(y) + \pi^2 v_3(y) + 2\pi v_2(y) + v_1(y) - 2\pi^2 v_1(y)v_2(y) - 2\pi v_1^2(y) = 0.$$

On the boundary, we have $v_i\big|_{y=0,1} = 0$.

We solve the first equation to get

$$v_1(y) = A \sin \pi y. \qquad (2.72)$$

Here A is a constant. The second equation then reads

$$v_2''(y) + \pi^2 v_2(y) = \pi^2 A^2 \sin^2 \pi y - 2\pi A \sin \pi y. \qquad (2.73)$$

Multiplying both sides by $\sin \pi y$ and integrating from 0 to 1, we have

$$\int_0^1 \left[v_2''(y) + \pi^2 v_2(y)\right] \sin \pi y \, dy = \pi^2 A^2 \times \frac{4}{3\pi} - 2\pi A \times \frac{1}{2}. \qquad (2.74)$$

After integration by parts, we see that

$$\int_0^1 v_2''(y) \sin \pi y \, dy = v_2'(y) \sin \pi y \Big|_{y=0}^1 - \pi \int_0^1 v_2'(y) \cos \pi y \, dy$$
$$= -\pi v_2(y) \cos \pi y \Big|_{y=0}^1 - \pi^2 \int_0^1 v_2(y) \sin \pi y \, dy. \qquad (2.75)$$

Therefore, the left hand side of (2.74) equals to 0. This gives $A = 0$ or $3/4$. The first one gives the trivial equilibrium (or $v(y) = \varepsilon^2 v_2(y) + \cdots$). We see that a necessary condition for the existence of the solution $v_2(y)$ determines the coefficient A.

One may ask whether or not this choice of A guarantees the existence of $v_2(y)$. To answer this question, we refer to Fredholm's alternative in functional analysis. As a matter of fact, we know that for an n-th order linear algebraic system $Ax = b$, it is necessary and sufficient for the existence of a solution x that b is orthogonal to the nullspace of A^T. By nullspace, we mean $\mathcal{N}(A^T) = \{y|\, A^T y = 0\}$.

To see this, we take any $y \in \mathcal{N}(A^T)$. This gives $y^T A = 0$. Multiplying y^T to $Ax = b$, we obtain $y^T b = y^T A x = 0$.

Therefore, if A is not singular, then for any b, there exists a unique solution $x = A^{-1}b$. If A is singular with rank k, then there exist $n - k$ linearly independent vectors $x_1, \cdots, x_{n-k}$, which span $\mathcal{N}(A)$, and $y_1, \cdots, y_{n-k}$, which span $\mathcal{N}(A^T)$. Fredholm's alternative asserts that if and only if we have $y_i^T b = 0, \quad i = 1, \cdots, n-k$, the system is solvable. A solution takes the form of $x = x^* + \sum_{i=1}^{n-k} C_i x_i$, where x^* is a special solution, and C_i are arbitrary coefficients.

In a Hilbert space H (complete inner product space), we define an inner product satisfying $(x, y) = (y, x)$. Consider an equation for an operator A and a point b, $Ax = b$. The adjoint operator B of A is defined by $(y, Ax) = (By, x), \forall x, y \in H$. Fredholm's alternative asserts that the equation has a solution if and only if $b \perp \mathcal{N}(B)$.

In case of a linear algebraic system, it may be readily shown that the adjoint operator of A is its transpose. For the ODE system on $v_2(y)$, we define an inner product by $(w, v) = \int_0^1 w(y)v(y)dy$, and the Hilbert space

$$H = \{w | \ (w, w) < +\infty, w(0) = w(1) = 0\}. \tag{2.76}$$

Consider the operator

$$D = \frac{d^2}{dy^2} + \pi^2. \tag{2.77}$$

We compute its adjoint operator as follows,

$$\begin{aligned}(w, Dv) &= \int_0^1 w(v'' + \pi^2 v)dy \\ &= \int_0^1 wv'' + \pi^2 wv \, dy \\ &= \int_0^1 w''v + \pi^2 wv \, dy + (-w'v + wv')\Big|_0^1 \\ &= \int_0^1 v(w'' + \pi^2 w)dy \\ &= (Dw, v).\end{aligned} \tag{2.78}$$

So the adjoint operator of D is itself. Hence it is a self-adjoint operator.

The nullspace of D is the set of all functions in the form of $\sin \pi y$, so Fredholm's alternative guarantees the solvability of the $v_2(y)$ equation under the choice of $A = 0, 3/4$.

Finally we make a remark on the perturbation method. The perturbation method presented above takes a formal expansion. An appropriate ansatz of the solution form is crucial. Experiences and understanding on the underlying system usually help specifying such an ansatz. After obtaining a solution in the series form via this approach, one may prove the series converges and the infinite sum is a solution to the original differential equation.

2.5 Traveling waves

Besides steady states, traveling waves are also special solutions that allow us to reduce the original reaction-diffusion system to an ODE system.

The linear wave equation describes wave propagation in one space dimension,

$$u_{tt} - c^2 u_{xx} = 0. \tag{2.79}$$

The D'Alambert formula gives a general solution as follows,

$$u(x,t) = f(x - ct) + g(x + ct). \tag{2.80}$$

Here functions f and g are determined by the initial data. They correspond to the right-going wave and the left-going wave, respectively.

A traveling wave in a reaction-diffusion equation

$$u_t = u_{xx} + f(u) \tag{2.81}$$

takes the form of $u(x,t) = u_T(\xi)$, with $\xi = x - ct$. This corresponds to a wave propagating from left to right at a constant velocity c. A traveling wave maintains its shape during propagation. As we shall see in a while, it exists only for a specific speed. Different traveling waves propagate in different speed and assume different profiles. Consequently, the investigation of a traveling wave has two coupled issues, the propagation speed c, and the wave profile $u_T(\xi)$.

To find a traveling wave, we substitute $u(x,t) = u_T(\xi)$ into the equation, and obtain a traveling wave equation

$$u_T'' + c u_T' + f(u_T) = 0. \tag{2.82}$$

This is a second order autonomous ODE. In most applications, we are interested only in a solution that remains finite, when ξ tends to infinity. There are two typical cases, namely, a heteroclinic/homoclinic orbit or a periodic orbit (closed orbit).

We illustrate by an example of Allen-Cahn type of reaction-diffusion equation

$$u_t = u_{xx} + u + u^2 - u^3. \tag{2.83}$$

See Fig. 2.6.

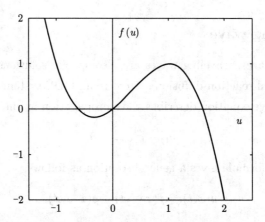

Figure 2.6 Reaction term in an Allen-Cahn type of equation.

The traveling wave equation is

$$u_T'' + c u_T' + u_T + u_T^2 - u_T^3 = 0. \tag{2.84}$$

For ease of presentation, we omit the subscript. We rewrite it into a system as

$$\begin{cases} u' = v, \\ v' = -cv - u - u^2 + u^3. \end{cases} \tag{2.85}$$

The critical points are $(u_i, 0)$ with $u_1 = \left(1 - \sqrt{5}\right)/2$, $u_2 = 0$, $u_3 = \left(1 + \sqrt{5}\right)/2$. The Jacobian matrix at these points is

$$\begin{bmatrix} 0 & 1 \\ -1 - 2u_i + 3u_i^2 & -c \end{bmatrix}. \tag{2.86}$$

The eigenvalues satisfy, for $i = 1, 2, 3$, respectively,

$$\begin{aligned} \lambda^2 + c\lambda - (5 - \sqrt{5})/2 &= 0; \\ \lambda^2 + c\lambda + 1 &= 0; \\ \lambda^2 + c\lambda - (5 + \sqrt{5})/2 &= 0. \end{aligned} \tag{2.87}$$

So, when $c = 0$, u_1 and u_3 are saddles, whereas u_2 is a center by virtue of symmetry. We emphasize here that a saddle is unstable only in the sense of the traveling wave variable ξ.

On the other hand, for an observable traveling wave, it should be stable in the sense of time evolution. Accordingly, the end-states of such a wave must be stable. More precisely, let us consider a traveling wave that connects two constant states, namely $\lim_{\xi \to \pm\infty} u(\xi) = u_\pm$. This implies that $\lim_{\xi \to \pm\infty} u'(\xi) = \lim_{\xi \to \pm\infty} u''(\xi) = 0$. Therefore, $u_\pm$ must be among u_1, u_2, u_3. Far away from the transition region, e.g., for $\xi \gg 0$, we essentially have a constant state u_+. Putting a small perturbation around this constant state, we should have a stable profile. Otherwise the traveling wave could not sustain.

Now we consider
$$u_t = u + u^2 - u^3. \tag{2.88}$$

It is readily shown that u_1 and u_3 are stable in terms of time evolution, and u_2 is unstable for this evolution problem without the diffusion term u_{xx}. So, we seek for a heteroclinic orbit in the traveling wave equation (2.84) that connects u_1 toward u_3. We remark that there actually exist a pair of such orbits due to symmetry.

The traveling wave equation at $c = 0$ gives a first integral
$$v^2 + u^2 + \frac{2u^3}{3} - \frac{u^4}{2} = C. \tag{2.89}$$

In particular, the trajectory passing through u_1 cannot reach u_3, because
$$u_1^2 + \frac{2u_1^3}{3} - \frac{u_1^4}{2} = \frac{4 + 5u_1}{6} < \frac{4 + 5u_3}{6} = u_3^2 + \frac{2u_3^3}{3} - \frac{u_3^4}{2}. \tag{2.90}$$

This is illustrated in subplot (a) of Fig. 2.7.

At $c < 0$, we observe that the traveling wave equation gives
$$\frac{dv}{du} = -c + \frac{-u - u^2 + u^3}{v}. \tag{2.91}$$

By comparison principle, we know that a smaller c gives a higher trajectory. Meanwhile, when $c \to -\infty$, it is obvious that the trajectory through $(u_1, 0)$ goes well above $(u_3, 0)$. Therefore, with the continuous dependency on parameter for

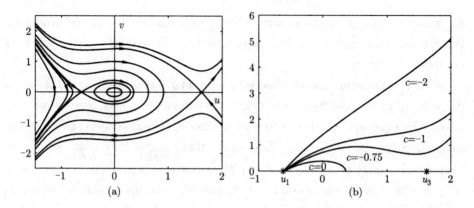

Figure 2.7 Traveling waves in an Allen-Cahn type of equation: (a) trajectories at $c = 0$; (b) trajectories from $(u_1, 0)$ for various c.

ODE solution, there exists a unique $c = c_0 < 0$, for which a heteroclinic orbit connects $(u_1, 0)$ and $(u_3, 0)$. This is the traveling wave.

We remark that there is another type of wave that also travels at a constant speed. This is related to a closed orbit of the corresponding traveling wave equation. Particular examples correspond to the standing wave ($c = 0$) as the homoclinic orbit and periodic orbits in the left-subplot of Fig. 2.7.

2.6 Burgers' equation and Cole-Hopf transform

The Navier-Stokes equations

$$\begin{cases} \rho_t + (\rho u)_x = 0, \\ (\rho u)_t + (\rho u^2 + p)_x = (\nu \rho u_x)_x \end{cases} \tag{2.92}$$

describe the dynamics of a fluid with density $\rho(x, t)$ and velocity $u(x, t)$. The pressure $p(\rho)$ is given by a certain equation of state as a function of density, and ν is a positive kinematic viscosity. The momentum equation may be easily rewritten as

$$\rho u_t + \rho u u_x + p_x = (\nu \rho u_x)_x. \tag{2.93}$$

This reduces to the following Burgers' equation when the density is a constant,

$$u_t + u u_x = \nu u_{xx}. \tag{2.94}$$

Burgers derived this equation in 1948 to study the nonlinear interaction between the convection term uu_x and the diffusion term νu_{xx}. He aimed at an understanding of turbulence, which turned out to be not adequate.

We rewrite Burgers' equation as follows,

$$u_t + \left(\frac{u^2}{2} - \nu u_x\right)_x = 0. \tag{2.95}$$

This is a convection-diffusion equation, instead of a reaction-diffusion equation. With this equation we illustrate a very special way to treat a nonlinear equation through a transform. This type of treatment rarely exists. But when there is a transform that changes a nonlinear problem into a linear one, we may obtain an exact analytical expression for the solutions. The explicit solutions are expected to considerably deepen our understanding for that type of nonlinear systems. We remark that the Cole-Hopf transform was invented by Cole and Hopf independently.

First, we observe that the divergence form of the equation may be regarded as a compatibility condition for a function ψ.

$$\psi_x = u, \quad -\psi_t = \frac{u^2}{2} - \nu u_x. \tag{2.96}$$

Substituting the first equation into the second one, we obtain

$$\psi_t = \nu \psi_{xx} - \frac{\psi_x^2}{2}. \tag{2.97}$$

If the temporal derivative term is regarded as a source term, we follow the standard transform in treating a nonlinear ODE $\psi = -2\nu \ln \varphi$, and express u in terms of φ. This gives the Cole-Hopf transform

$$u = -2\nu \frac{\varphi_x}{\varphi}. \tag{2.98}$$

Now we compute the derivatives of ψ.

$$\psi_t = -2\nu \frac{\varphi_t}{\varphi}, \quad \psi_{xx} = 2\nu \frac{\varphi_x^2}{\varphi^2} - 2\nu \frac{\varphi_x}{\varphi}. \tag{2.99}$$

The ψ-equation becomes

$$-2\nu \frac{\varphi_t}{\varphi} = \nu \left(2\nu \frac{\varphi_x^2}{\varphi^2} - 2\nu \frac{\varphi_{xx}}{\varphi}\right) - \frac{1}{2}\left(2\nu \frac{\varphi_x}{\varphi}\right)^2. \tag{2.100}$$

After some manipulations, we obtain the heat equation

$$\varphi_t = \nu \varphi_{xx}. \tag{2.101}$$

To obtain an explicit expression for $u(x,t)$, we consider Cauchy data $u(x,0) = u_0(x)$. Corresponding initial data for φ is solved from

$$u_0(x) = -2\nu \frac{\varphi_x(x,0)}{\varphi(x,0)}. \tag{2.102}$$

By integration, we have

$$\varphi(x,0) = e^{-\frac{1}{2\nu}\int_0^x u_0(\alpha)d\alpha}. \tag{2.103}$$

The solution to the heat equation is

$$\varphi(x,t) = \frac{1}{2\sqrt{\pi \nu t}} \int_{-\infty}^{+\infty} e^{-\frac{1}{2\nu}\left[\int_0^y u_0(\alpha)d\alpha + \frac{(x-y)^2}{2t}\right]} dy. \tag{2.104}$$

The derivative is

$$\varphi_x(x,t) = -\frac{1}{4\nu\sqrt{\pi \nu t}} \int_{-\infty}^{+\infty} \frac{x-y}{t} e^{-\frac{1}{2\nu}\left[\int_0^y u_0(\alpha)d\alpha + \frac{(x-y)^2}{2t}\right]} dy. \tag{2.105}$$

In summary, the solution to Burgers' equation is

$$u(x,t) = \frac{\int_{-\infty}^{+\infty} \frac{x-y}{t} e^{-\frac{1}{2\nu}\left[\int_0^y u_0(\alpha)d\alpha + \frac{(x-y)^2}{2t}\right]} dy}{\int_{-\infty}^{+\infty} e^{-\frac{1}{2\nu}\left[\int_0^y u_0(\alpha)d\alpha + \frac{(x-y)^2}{2t}\right]} dy}. \tag{2.106}$$

2.7 Evolutionary Duffing equation

In a reaction-diffusion equation, diffusion usually smooths things out, and drives the system to spatially homogeneous equilibrium. On the other hand, the reaction term may serve as the stimulus when the system lies at equilibrium. The interplay

Chapter 2 Reaction-Diffusion Systems

between these two mechanisms generates various scenario under different range of parameters and with different initial boundary data.

In this section, we explore the evolutionary Duffing equation (2.52) mainly through numerical simulations.

For simplicity, we consider a spatially periodic problem. After rescaling, we set the spatial period as 2π and recast (2.52) into the following form with a controlling parameter k,

$$u_t = k^2 u - u^3 + u_{xx}. \qquad (2.107)$$

We first discretize the interval $(0, 2\pi]$ by a uniform mesh with mesh size $\Delta x = 2\pi/128$. We take a time step size $\Delta t = 0.001$. The numerical solution is then described by a collection of values at the grid points $u_i^n = u(i\Delta x, n\Delta t)$. While the initial data is chosen as $u_i^0 = u(i\Delta x, 0)$, we need to design a scheme to compute u_i^{n+1}'s from u_i^n's.

We take a central difference for u_{xx} as follows,

$$u_{xx}(i\Delta x, n\Delta t) \sim \frac{u_{i-1}^n - 2u_i^n + u_{i+1}^n}{(\Delta x)^2}. \qquad (2.108)$$

Due to periodicity, we evaluate this central difference at the boundary points as follows,

$$u_{xx}(\Delta x, n\Delta t) \sim \frac{u_{128}^n - 2u_1^n + u_2^n}{(\Delta x)^2}, \qquad (2.109)$$

$$u_{xx}(128\Delta x, n\Delta t) \sim \frac{u_{127}^n - 2u_{128}^n + u_1^n}{(\Delta x)^2}. \qquad (2.110)$$

A forward Euler discretization is adopted for the time derivative. Hence we obtain an explicit scheme,

$$u_i^{n+1} = u_i^n + \Delta t \left(k^2 u_i - u_i^3 + \frac{u_{i-1}^n - 2u_i^n + u_{i+1}^n}{(\Delta x)^2} \right). \qquad (2.111)$$

We start with an investigation with $k = 0$ and initial data $u_0(x) = 0.1 \sin 2x + 0.01$. The Matlab code is as follows.

```
k=1;

nx=128; % Number of spatial grid points dx=2*pi/nx;
```

```
x=[0:dx:2*pi-dx];
nt=150; % Number of big time steps

dt=0.001; % time step size
nt1=100; % number of time steps in a big step

% Initial data
u(:,1)=sin(2*x)*0.1+0.001;
plot(u(:,1));axis([0 nx -k-1,k+1]); utmp=u(:,1);

for j1=1:nt;
   for j2=1:nt1;
      du(1:nx,1)=0;
      du(2:nx-1,1)=(utmp(1:nx-2)+utmp(3:nx)-2*utmp(2:nx-1))/dx/dx;
      du(1,1)=(utmp(2)+utmp(nx)-2*utmp(1))/dx/dx;
      du(nx,1)=(utmp(nx-1)+utmp(1)-2*utmp(nx))/dx/dx;
      du=du+k*k*utmp-utmp^3;
      utmp=utmp+dt*du;
   end;
   u(:,j1+1)=utmp;
   plot(u(:,j1+1));axis([0 nx -k-1,k+1]);pause(0.1);
end;
```

We depict the evolution in Fig. 2.8 (a). The spatial variation diminishes quickly, and at around $t = 8$, the spatially homogeneous state quickly moves up to $u_+ = 1$.

When $k = 4$, the picture is different. There first develops a two-hump solution, which looks quite stable. But this two-hump solution moves rapidly to the homogeneous state $u_+ = 4$ at around $t = 12$. See Fig. 2.8 (b).

In general the system converges to an equilibrium state, if the mean of initial data is none-zero.

In the following we simulate the same problem, yet with zero-mean initial data $u_0(x) = 0.1 \sin x$. From Fig. 2.9 (a), we observe that the system quickly converges

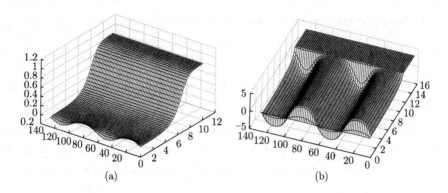

Figure 2.8 Evolutionary Duffing equation with none zero-mean initial data: (a) $k = 1$; (b) $k = 4$.

toward the one-hump solution. On the other hand, if we take initial data $u_0(x) = 0.1 \sin 2x$, the system seems converge to the zero equilibrium in Fig. 2.9 (b).

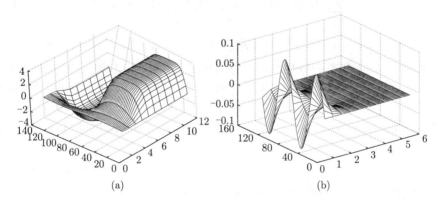

Figure 2.9 Evolutionary Duffing equation with $k = 1.2$ and zero-mean initial data: (a) $0.1 \sin x$; (b) $0.1 \sin 2x$.

The stationary solutions are governed by the Duffing equation. For the Cauchy problem and the Neumann boundary value problem, it may be readily shown that only the constant solution $u_\pm = \pm k$ is temporally stable. The zero equilibrium is unstable.

The dispersion relation for the linearized equation around zero equilibrium

$$\tilde{u}_t = k^2 \tilde{u} + \tilde{u}_{xx} \tag{2.112}$$

is $\lambda = k^2 - p^2$ for a Fourier mode $e^{\lambda t + ipx}$. If the spatially periodic system maintains the period of initial data, then the 2π-periodic perturbation in the form of $\sin x$ is unstable for $k = 1.2$, whereas the π-periodic perturbation in the form of $\sin 2x$ is stable. Nevertheless, noticing that numerical round-off error always exists, we expect eventually the system will leave the zero equilibrium in a long run. So, we confine ourselves to the explorations with zero mean, namely, $\int_0^{2\pi} u(x,t)\mathrm{d}x = 0$.

Correspondingly, we adopt another type of numerical method to ensure the zero-mean. We compute by the Fourier pseudo-spectral method with integrating factors and a standard fourth order Runge-Kutta scheme. More precisely, we approximate a solution by sine series to force the zero mean value,

$$u(x,t) = \sum_{p=1}^{N-1} A_p(t) \sin px, \quad u^3(x,t) = \sum_{p=1}^{N-1} B_p(t) \sin px. \tag{2.113}$$

Here B_p is a function of $(A_1, \cdots, A_{N-1})$ through sine transform and inverse sine transform. Taking new variables $C_p = \exp\left[-(k^2 - p^2)t\right] A_p$, we obtain a dynamical system for C_p. A standard fourth-order Runge-Kutta scheme is employed for time integration.

With this zero-mean constraint, the equilibria $u_\pm$ are excluded. Furthermore, the zero-equilibrium is unstable for $k > 1$ as we take the computation within the interval $[0, 2\pi]$. On the other hand, there are non-constant patterns in spatially periodic problem of (2.107) with constraint. They correspond to closed orbits of the Duffing equation.

Following the discussions about the Duffing equation in Chapter 1, we observe that the period L of a closed orbit is related to its amplitude U_0 by

$$L = \frac{4}{k} \int_0^1 \left[1 - s^2 - U_0^2 \left(1 - s^4\right)/2k^2\right]^{-1/2} \mathrm{d}s. \tag{2.114}$$

This can be reformulated in terms of elliptic functions. At a fixed k, as U_0/k varies from 0 to 1, the period varies monotonously from $2\pi/k$ to ∞.

By virtue of the invariance in (2.107), we may study the spatially periodic problem with a fixed period 2π. Then there are finite many candidates for steady state, namely, $u_n(x)$ with a period $2\pi/n$, for $n = 1, \cdots, [k] - 1$. We shall investigate how the system selects one among them as the time-asymptotic solution, in particular, how this pattern selection depends on initial profile and the controlling parameter.

For $k > K_1 = 1$, a one-hump pattern forms, and turns out to be stable. For instance, at $k = 5$, we calculate for initial profile

$$u(x, 0) = \begin{cases} e^{-10(x-\pi/2)^2}, & 0 < x < \pi, \\ -e^{-10(x-3\pi/2)^2}, & \pi < x < 2\pi. \end{cases} \quad (2.115)$$

The system quickly converges to the one-hump pattern. At $t = 0.5$, it already looks stable. Checking with the data files, there undergoes slight adjustment afterwards. See Fig. 2.10. The amplitude is $U_0 = 4.9996999$, for which the corresponding period is away from 2π by 6.628×10^{-7}.

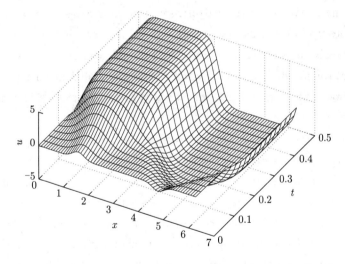

Figure 2.10 Evolution towards a one-hump pattern at $k = 5$.

As we shall see in a moment, multi-hump patterns exist. Rigorous analysis with centered manifold theory shows that patterns with interface(s) may move slowly and persist for exponentially long time. This exponential stability of unstable patterns indicates the difficulty in numerical study. They are meta-stable, and an extremely small amount of perturbation may drive the system towards the one-hump pattern eventually. This is illustrated by an example at $k = 5$. We start with the two-hump pattern perturbed by $10^{-7} \sin x$.

The switching to the one-hump pattern consists of two steps. In the first step, this anti-symmetry in the two-hump pattern breaks. It takes very long time to reach an observable level. In the second step, the system switches to the one-hump pattern suddenly around $t = 960$. The details can be found in Fig. 2.11. The growth of A_1 is an indicator for the pattern switching.

Here we make a few remarks. First, at a controlling parameter k when no meta-stable pattern exists, the choice of initial profile is not important. Numerical experiments show that the system reaches quickly the same one-hump pattern for all initial profiles we have used. Secondly, if we supply no perturbation initially, the perturbation comes mainly from round-off errors.

Now we take initial data containing only the n-th mode, e.g., starting with $u(x, 0) = \sin nx$. We study the pattern selection at increasing k, and determine critical values up to the accuracy of 0.01.

Theoretically, an exact solution to (2.107) should keep $2\pi/n$-periodic if initially so. However, round-off errors are inevitable, containing all frequencies. As a result, we shall numerically obtain a pattern only if it sustains perturbations at the level of round-off errors.

When $n = 2$, the period is $L = \pi$. Numerically the two-hump pattern starts to be selected at $k \geqslant K_2 = 4.14$, as shown in Fig. 2.12.

We search for stable three-hump patterns, but we have not find it in our computations for $k \in [0, 20]$. At $k = 20$, it is seen that the system moves up to a three-hump pattern at $t = 1$. However, we observe a linear growth of the first mode for time up to 100, continued by a nonlinear increment. This increment triggers a switching to a one-hump pattern at time between 8000 to 9000. We regard this growth a definite indication of the instability for the three-hump pattern. It is worth noting that the perturbation comes from approximation errors as well as round-off errors. With more modes, the growth of the first mode is slower. See Fig. 2.13.

At $n = 4$, we obtain a stable four-hump pattern for $k \geqslant 8.38$. See Fig. 2.14.

Similar to the case of $n = 3$, we have not obtained five-, six-, or seven-hump

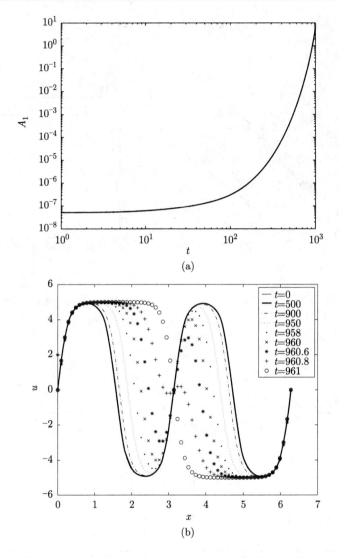

Figure 2.11 Perturbed problem for two-hump pattern at $k = 5$: (a) growth of the amplitude of $\sin x$; (b) the drastic change around $t = 960$.

patterns for $k \leqslant 20$. The case for $n = 6$ is quite interesting because it is a multiple of 2. The system selects the one-hump pattern for $k < 12.89$, and the two hump pattern for $k \geqslant 12.89$.

When $k \geqslant 17.86$, an eight-hump pattern is reached.

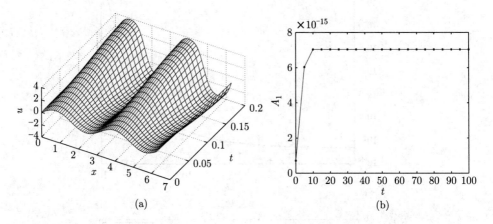

Figure 2.12   Two-hump pattern at $k = 4.14$: (a) evolution at early stage; (b) the amplitude of $\sin x$.

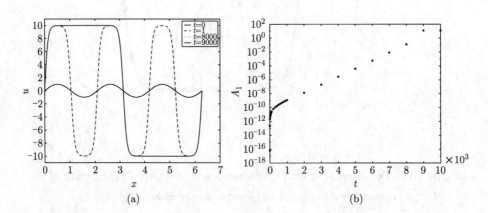

Figure 2.13   Evolution at $k = 10$ with initial data $u(x, 0) = \sin 3x$: (a) snapshots; (b) the amplitude of $\sin x$.

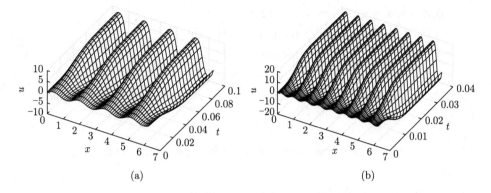

Figure 2.14 More humps: (a) four-hump pattern at $k = 8.39$; (b) eight-hump pattern at $k = 17.86$.

Assignments

1. What is the solution if we prescribe Dirichlet boundary conditions on all boundaries for the Laplace equation? Take $u(0,y) = u(a,y) = u(x,0) = u(x,b) = 0$.

2. A function $f(x,y)$ that solves the Laplace equation is called a harmonic function. Write some such functions.

3. Find the steady states of the Fisher equation

$$u_t = u_{xx} + u(1-u). \tag{2.116}$$

4. Find the first integral and then investigate the ODE system

$$\begin{aligned} D_1 p_{xx} - pq &= 0, \\ D_2 q_{xx} + pq &= 0. \end{aligned} \tag{2.117}$$

5. To your knowledge, which type of functions can be expressed as a superposition of trigonometric functions? For instance, $f(x) = \sum_{k=0}^{\infty} C_k \cos k\pi x + \sum_{k=1}^{\infty} D_k \sin k\pi x$, or $f(x) = \int F(\omega) e^{i\omega x} d\omega$.

6. Check that the L^2 norm is a norm, i.e.,

(a) It is non-negative for any function;
(b) It is zero if and only if the function is zero;
(c) $\|\lambda f(x)\|_2 = |\lambda|\, \|f(x)\|_2$ for any $\lambda \in \mathbb{R}$;
(d) $\|f(x) + g(x)\|_2 \leqslant \|f(x)\|_2 + \|g(x)\|_2$.

7. Study the linear stability of $u(x,t) = 0$ in an evolutionary Duffing equation
$$u_t = u - u^3 + u_{xx}. \tag{2.118}$$

8. For the competition equations, construct invariant domains under other situations.

9. For the Belousov-Zhabotinskii reaction system
$$\begin{cases} \xi_t = \xi_{xx} + \xi + \eta - \xi^2 - \xi\eta, \\ \eta_t = \eta_{xx} - \eta + 2\rho - \xi\eta, \\ \rho_t = \rho_{xx} + \xi - \rho, \end{cases} \tag{2.119}$$
show that $[0,2] \times [0,7] \times [0,3]$ forms an invariant domain.

10. In the perturbation study of the Fisher equation, find the solution $v_2(y)$ by using a similar argument as that for $v_1(y)$, namely, find the amplitude through the equation for $v_3(y)$.

11. For the example of perturbation, we may take a discretization with step size $\Delta y = 1/N$. Denoting $v_2^n = v_2(n\Delta y)$, $n = 0, \cdots, N$, we have equations for $n = 1, \cdots, N-1$,
$$v_2^{n+1} + (\pi^2(\Delta y)^2 - 2)v_2^n + v_2^{n-1} = \left(\pi^2 A^2 \sin^2(\pi n \Delta y) - 2\pi A \sin(\pi n \Delta y)\right)(\Delta y)^2. \tag{2.120}$$
The boundary conditions are $v_2^0 = v_2^N = 0$. With the notation
$$V_2 = \begin{bmatrix} v_2^1 \\ \vdots \\ v_2^{N-1} \end{bmatrix},$$
show that the resulting linear algebraic system has a unique solution for any finite N. Discuss the situation when $N \to \infty$.

12. For a reaction-diffusion system
$$u_t = u_{xx} + u(1-u)(u-a), \tag{2.121}$$

with $0 < a < 1/2$, show that $u_T(\xi) = 1/\left[1 + \exp(-\xi/\sqrt{2})\right]$ is a traveling wave with $c = -1/\sqrt{2} + \sqrt{2}a$. Find an explicit solution for the system (namely, only for a certain specific c)

$$\begin{cases} u' = v, \\ v' = -cv - u - u^2 + u^3. \end{cases} \quad (2.122)$$

13. Show that $\varphi(x,t) = 1 + \sqrt{\dfrac{\tau}{t}} e^{-\frac{x^2}{4\nu t}}$ is a solution to the heat equation. Show that the corresponding solution to Burgers' equation is

$$u(x,t) = \dfrac{x}{t\left(1 + \sqrt{\dfrac{t}{\tau}} e^{\frac{x^2}{4\nu t}}\right)}. \quad (2.123)$$

14. Noticing that in the previous expression, we have $u(x,t) > 0$ for $x > 0$, $u(0,t) = 0$, and $u(x,t) < 0$ for $x < 0$. We divide the wave profile accordingly into a positive phase and a negative phase. With $\nu = 1, \tau = 1$, please plot the solution at time $t = 0.01, 0.1, 1, 2, 10, 100$ and describe the evolution.

Chapter 3 Elliptic Equations

3.1 Sobolev spaces

An elliptic equation usually governs the stationary or asymptotic profile of a parabolic equation. Over the last century, abundant remarkable theories for elliptic equations have been developed. We only sketch a simple part of these developments.

For a mathematician, some fundamental issues in the qualitative study of a differential equation include existence, uniqueness and stability. We have explored these issues for ODE's. For partial differential equations, these are very challenging topics. In general, classical solutions may not exist. By a classical solution, we mean a function that is differentiable to the required order, and satisfies the partial differential equation. Instead of finding such a solution, one has to lower the expectation to a weak solution, which means to satisfy the equation in a weak sense, namely, in a certain integral sense. For one thing, two functions that are different from each other only over a very small subset of the spatial domain are regarded identical. This way for finding a solution may be rephrased that we search for solutions in a function space which is bigger than C^2. A minimization sequence or an iteration procedure usually provides the existence of a weak solution(s). On the other hand, the function space should not be too big, otherwise the uniqueness can not be obtained. Therefore the selection of a proper function space is vital in such explorations. To investigate elliptic equations, the most appropriate function spaces fall into the category of Sobolev spaces. Here we shall confine ourselves to a special sub-category of spaces, namely, the Hilbert spaces.

We consider $\Omega \in \mathbb{R}^n$ a bounded open set with piecewise smooth boundary. Moreover, a cone condition is assumed in most applications to elliptic partial differential equations. That is, at any point of $\partial\Omega$, a cone with a positive inner angle is locally contained within Ω.

A Hilbert space is a complete space with inner product. Depending on the inner product defined, we have a sequence of Hilbert spaces for functions.

We start with $H^0(\Omega)$, which is actually $L_2(\Omega)$. We define an inner product

$$(u, v)_0 = \int_\Omega u(x) v(x) \mathrm{d}x. \tag{3.1}$$

This leads to an L_2 norm $\| u \|_0 = \sqrt{(u,u)_0} = \sqrt{\int_\Omega u(x)^2 \mathrm{d}x}$. The space $L_2(\Omega)$ is the collection of all functions with finite L^2-norm.

We notice that a function in $L_2(\Omega)$ may not be differentiable in general. So, we define a weak derivative instead. For this purpose, we define $C^\infty(\Omega)$ the smooth function space, and $C_0^\infty(\Omega)$ the subspace with each element taking a compact support. Because Ω is open, and a compact set in $\mathbb{R}$ must be bounded and closed, we know that a function in $C_0^\infty(\Omega)$ must vanish at the boundary.

For $u \in L^2$, we define its weak derivative $v = \partial^\alpha u \in L^2$, where α is a multiple index, if

$$(\phi, v)_0 = (-1)^{|\alpha|} (\partial^\alpha \phi, u)_0, \quad \forall \phi \in C_0^\infty(\Omega). \tag{3.2}$$

We further define an inner product and corresponding norm as follows,

$$(u, v)_m = \int_\Omega \sum_{|\alpha| \leqslant m} (\partial^\alpha u, \partial^\alpha v) \mathrm{d}x, \quad \| u \|_m = \sqrt{(u,u)_m}. \tag{3.3}$$

A semi-norm may be defined by

$$|u|_m = \sqrt{\sum_{|\alpha|=m} \| \partial^\alpha u \|_0^2}. \tag{3.4}$$

The Hilbert space is defined as $H^m(\Omega) = \{ u \in L^2(\Omega) | \; \| u \|_m < +\infty \}$. The completeness may be verified. It is possible to show that $C^\infty(\Omega) \cap H^m(\Omega)$ is dense in $H^m(\Omega)$, and $H^m(\Omega)$ is the completion of $C^\infty(\Omega) \cap H^m(\Omega)$ under the H^m norm.

In a similar way, we define $H_0^m(\Omega)$ the completion of $C_0^\infty(\Omega)$. There are two sequences of inclusion,

$$\begin{array}{ccccccc} L_2(\Omega) = H^0(\Omega) & \supset & H^1(\Omega) & \supset & H^2(\Omega) & \supset & \cdots \\ \| & & \cup & & \cup & & \\ H_0^0(\Omega) & \supset & H_0^1(\Omega) & \supset & H_0^2(\Omega) & \supset & \cdots. \end{array} \tag{3.5}$$

In particular, the norm and the seminorm are equivalent in $H_0^m(\Omega)$ by the following result based on the Poincare-Friedrichs inequality,

$$|v|_m \leqslant \|v\|_m \leqslant C|v|_m. \tag{3.6}$$

Here C depends on the size of Ω.

A key fact in Sobolev spaces is the compact imbedding. For $m > 0$ and Ω a Lipschitiz domain with cone condition, $H^{m+1}(\Omega) \hookrightarrow H^m(\Omega)$ is a compact imbedding, namely, a subset which is bounded in H^{m+1} is relatively compact in H^m. The compact imbedding facilitates theoretical studies of elliptic and parabolic partial differential equations, such as the existence and regularity of the solutions.

3.2 Variational formulation of second-order elliptic equations

A simple example of elliptic partial differential equation is the Laplace equation

$$\Delta u = u_{xx} + u_{yy} = 0, \quad (x,y) \in \Omega. \tag{3.7}$$

Notice that a certain boundary condition is necessary.

More generally, we may consider an elliptic operator

$$Lu := \sum_{i,k=1}^{d} a_{ik}(x,u) u_{x_i x_k}, \tag{3.8}$$

where the coefficient square matrix (a_{ik}) is positive definite. Moreover, if $\exists \alpha > 0, \forall x \in \Omega$, and u in a suitable function space, all the eigenvalues are no smaller than α, we call L uniformly elliptic. For such an operator, some features are as follows.

- **Maximum principle.**

  If $Lu = f \leqslant 0$, then u attains its maximum on $\partial \Omega$.

- **Comparison principle.**

  If two classical solutions u, v satisfy $Lu \leqslant Lv$ in Ω, and $u \geqslant v$ on $\partial \Omega$, then it holds that $u \geqslant v$ in Ω.

- **Continuous dependency on the boundary data.**
  For two solutions of $Lu_i = f$ ($i = 1, 2$) with different boundary data, it holds that $\sup_{x \in \Omega} |u_1(x) - u_2(x)| = \sup_{z \in \partial\Omega} |u_1(z) - u_2(z)|$.
- **Continuous dependency on the righthand side.**
  $\forall u \in C^2(\Omega) \cap C^0(\bar{\Omega})$, it holds that $|u(x)| \leqslant \sup_{\partial\Omega} |u(z)| + C \sup_{\partial\Omega} |Lu(z)|$.
- **Elliptic operator with Helmholtz term.**
  For operator $Lu = -\sum a_{ik}(x) u_{x_i x_k} + c(x) u$ with $c(x) \geqslant 0$, if $Lu \leqslant 0$, then $\sup_{x \in \Omega} u(x) \leqslant \max\{0, \sup_{\partial\Omega} u(z)\}$.

In the following, we shall consider an elliptic operator in divergence form $Lu = -\sum_{i,k} \partial_i (a_{ik} \partial_k u) + a_0 u$, with $(a_{ik}), a_0(x) \geqslant 0$. For this operator, we may take an associated bilinear form as follows,

$$a(u, v) = \int_\Omega \left(\sum_{i,k} a_{ik} \partial_i u \partial_k v + a_0 u v \right) dx. \tag{3.9}$$

As a classical solution does not exist in general, we consider a weak solution instead. We call $u \in H_0^1(\Omega)$ a weak solution of $Lu = f$ in Ω and $u = 0$ on $\partial\Omega$ if

$$a(u, v) = (f, v)_0, \quad \forall v \in H_0^1(\Omega). \tag{3.10}$$

We transform the partial differential equation into a minimization problem. The following theorem relates an equation with the minimization of a bilinear form. In a sense, it is similar to Fermat's theorem in Calculus, which states that an extremum point must be a critical point.

Theorem 3.2.1 (Characterization) *Let V be a linear space, $a : V \times V \to \mathbb{R}$ be a symmetric positive bilinear form, and $l : V \to \mathbb{R}$ be a linear functional denoted as $\langle l, v \rangle = l(v)$. Then*

$$J(v) = \frac{1}{2} a(v, v) - <l, v> \tag{3.11}$$

attains its minimum over V at u if and only if

$$a(u, v) = <l, v>, \quad \forall v \in V. \tag{3.12}$$

Moreover, there is at most one solution.

Proof. Take $u, v \in V, t \in \mathbb{R}$, we compute

$$
\begin{aligned}
J(u+tv) &= \frac{1}{2}a(u+tv, u+tv) - <l, u+tv> \\
&= J(u) + t[a(u,v) - <l,v>] + \frac{1}{2}t^2 a(v,v).
\end{aligned}
\tag{3.13}
$$

On one hand, if $a(u,v) = <l,v>, \forall v \in V$, then we have

$$
J(u+tv) = J(u) + \frac{1}{2}t^2 a(v,v) > J(u), \quad \forall v \in V \text{ and } v \neq 0.
\tag{3.14}
$$

On the other hand, if J attains minimum at u, then $\forall v$, we consider the function $t \longmapsto J(u+tv)$. By Fermat's theorem we have

$$
\frac{\mathrm{d}}{\mathrm{d}t} J(u+tv)|_{t=0} = a(u,v) - <l,v> = 0.
\tag{3.15}
$$

Finally, we prove the uniqueness. If two solutions satisfy $a(u_1, v) = <l, v> = a(u_2, v)$, then $J(u_1)$ and $J(u_2)$ are both minimum. So $J(u_1 + (u_2 - u_1)) > J(u_1)$ leads to a contradiction. □

The theorem implies that every classical solution of the boundary value problem

$$
\begin{cases}
-\sum_{i,k} \partial_i(a_{ik} \partial_k u) + a_0 u = f, & \text{in } \Omega, \\
u = 0, & \text{on } \partial\Omega,
\end{cases}
\tag{3.16}
$$

is a solution of the variational problem $v \in C^2(\Omega) \cap C^0(\bar{\Omega})$ with $v|_{\partial\Omega} = 0$.

$$
J(v) = \int_\Omega \left[\frac{1}{2} \sum a_{ik} \partial_i v \partial_k v + \frac{1}{2} a_0 v^2 - fv \right] \mathrm{d}x \to \min!.
\tag{3.17}
$$

The following Lax-Milgram theorem is key to the understanding of elliptic partial differential equations.

Theorem 3.2.2 (Lax-Milgram) *Let V be a closed convex set in a Hilbert space H, and $a : H \times H \to \mathbb{R}$ be an elliptic bilinear form. Then $\forall l \in H'$ (dual space of H), there is a unique $v \in V$ to the problem*

$$
J(v) = \frac{1}{2}a(v,v) - <l,v> \to \min!.
\tag{3.18}
$$

Proof. We claim that J is bounded from below. As a matter of fact, due to the ellipticity and the definition of dual space, we have

$$J(v) \geq \frac{1}{2}\alpha \parallel v \parallel^2 - \parallel l \parallel \parallel v \parallel = \frac{1}{2\alpha}(\alpha \parallel v \parallel - \parallel l \parallel)^2 - \frac{\parallel l \parallel^2}{2\alpha} \geq -\frac{\parallel l \parallel^2}{2\alpha}. \quad (3.19)$$

Therefore, there exists $\inf J(v) = C$. Let (v_n) be a minimizing sequence. We derive

$$\begin{aligned}
\alpha \parallel v_n - v_m \parallel^2 &\leq a(v_n - v_m, v_n - v_m) \\
&= 2a(v_n, v_n) + 2a(v_m, v_m) - a(v_n + v_m, v_n + v_m) \\
&= 4J(v_n) + 4J(v_m) - 8J\left(\frac{v_n + v_m}{2}\right) \\
&\leq 4J(v_n) + 4J(v_m) - 8C.
\end{aligned} \quad (3.20)$$

Here we have made usage of the convexity of V to derive that $\dfrac{v_n + v_m}{2} \in V$. The above term tends to 0 as $n, m \to \infty$. Now combining the facts that $\{v_n\}$ is a Cauchy sequence, H is complete, and V is closed, we conclude that

$$u = \lim_{n \to \infty} v_n \in V. \quad (3.21)$$

Furthermore, as J is continuous, $J(u) = \lim\limits_{n\to\infty} J(v_n) = C_1 = \inf\limits_{v \in V} J(v)$.

Next, we prove the uniqueness. To this end, assuming that u_1, u_2 are both solutions, we may construct a sequence $(u_1, u_2, u_1, u_2, \cdots)$. It is obviously a minimizing sequence. It then must be a Cauchy sequence, and hence $u_1 = u_2$. □

We remark that the difference between the characterization theorem and the Lax-Milgram theorem lies in the difference of the space. In the previous one, the whole Hilbert space is adopted. On the other hand, the Lax-Milgram theorem uses only a closed convex subset.

Theorem 3.2.3 (Existence) Let L be a uniformly elliptic operator, with $a_0, a_{ij} \in L_\infty(\Omega), a_0 \geq 0, f \in L_2(\Omega)$. The boundary value problem

$$\begin{cases} Lu = f, & \text{in } \Omega, \\ u = 0, & \text{on } \partial\Omega \end{cases} \quad (3.22)$$

admits a weak solution in $H_0^1(\Omega)$. It is a minimum of

$$\frac{1}{2}a(v, v) - (f, v)_0 \to \min! \quad \text{over} \quad H_0^1(\Omega). \quad (3.23)$$

Proof. It is possible to show that

$$\left|\sum_{i,k}\int a_{ik}\partial_i u \partial_k v dx\right| \leqslant C\sum_{i,k}\int |\partial_i u \partial_k v| dx \leqslant C|u|_1|v|_1. \qquad (3.24)$$

Furthermore, we have

$$\left|\int a_0 uv dx\right| \leqslant C\|u\|_0\|v\|_0. \qquad (3.25)$$

These lead to

$$a(u,v) \leqslant C\|u\|_1\|v\|_1. \qquad (3.26)$$

For any $v \in C^1$, it holds that

$$\sum_{i,k} a_{ik}\partial_i v \partial_k v \geqslant \alpha \sum_i (\partial_i v)^2. \qquad (3.27)$$

Therefore, for any $v \in H^1$, it holds that

$$a(v,v) \geqslant \alpha |v|_1^2. \qquad (3.28)$$

From the Poincare-Friedrichs inequality, we know that

$$|v|_1 \sim \|v\|_1. \qquad (3.29)$$

Combining (3.26)–(3.29), we find that a is an elliptic bilinear form on $H_0^1(\Omega)$. Noticing that $f \in L^2 \subset H'$, we conclude the existence and uniqueness of the weak solution from the Lax-Milgram theorem. □

We remark that the above results may be extended to non-homogeneous Dirichlet boundary value problem. Consider

$$\begin{cases} Lu = f, & \text{in } \Omega, \\ u = g, & \text{on } \partial\Omega. \end{cases} \qquad (3.30)$$

We assume $\exists u_0$, such that Lu_0 exists, and $u_0|_{\partial\Omega} = g$. Now let $w = u - u_0$, then w solves a homogeneous boundary value problem

$$\begin{cases} Lw = f - Lu_0 \equiv f_1, & \text{in } \Omega, \\ w = 0, & \text{on } \partial\Omega. \end{cases} \qquad (3.31)$$

3.3 Neumann boundary value problem

Suppose that $\mathbf{n} = (n_i)$ is the outer normal on the boundary $\Gamma = \partial \Omega$. The Neumann boundary value problem refers to the following setting,

$$\begin{cases} Lu = f, & \text{in } \Omega, \\ \sum_{i,k} n_i a_{ik} \partial_k u = g, & \text{on } \Gamma. \end{cases} \quad (3.32)$$

It is obvious that the solution is not in $H_0^1(\Omega)$ in general. In fact, if u is a solution, so is $u + C$ with C a constant. The suitable function space is $H^1(\Omega)$.

Ellipticity in $H^1(\Omega)$ requires $a_{ik} \geq \alpha \geq 0$ and $a_0 \geq \alpha$ in Ω. Consequently, we see $\forall v \in H^1(\Omega)$,

$$a(v,v) = \int_\Omega \left[\sum_{i,k} a_{ik} \partial_i u \partial_k v + a_0 v^2 \right] dx \geq \alpha |v|_1^2 + \alpha \|v\|^2 = \alpha \|v\|_1^2. \quad (3.33)$$

Using a certain trace theorem, it may be proved that $< l, v > = \int_\Omega f v dx + \int_\Gamma g v ds$ defines a bounded linear functional with $f, g \in L_2(\Gamma)$. Again, the boundary value problem may be transformed to a variational problem.

Theorem 3.3.1 *Suppose that Ω is a bounded domain with piecewise smooth boundary, and satisfying the cone condition, then the variational problem*

$$\frac{1}{2} a(v,v) - (f,v)_{0,\Omega} - (g,v)_{0,\Gamma} \to \min! \quad (3.34)$$

has a unique solution $u \in H^1(\Omega)$. Moreover, $u \in C^2(\Omega) \cap C^1(\bar{\Omega})$ if classical solution exists for

$$\begin{cases} Lu = f, & \text{in } \Omega, \\ \sum_{i,k} n_i a_{ik} \partial_k u = g, & \text{on } \Gamma. \end{cases} \quad (3.35)$$

As an example, the Poisson equation

$$\begin{cases} -\Delta u = f, & \text{in } \Omega, \\ \dfrac{\partial u}{\partial \mathbf{n}} = g, & \text{on } \Gamma \end{cases} \quad (3.36)$$

admits a unique solution up to a constant. If we restrict the solution to $V = \{ v \in H^1(\Omega), \int_\Omega v dx = 0 \}$, then uniqueness is obtained. In fact, the Poincare-Friedrichs inequality implies that $a(u,v) = \int_\Omega \nabla u \cdot \nabla v \, dx$ is elliptic in V. We remark that a compatibility condition is required due to the Gauss theorem, namely, $\int_\Omega f \, dx + \int_\Gamma g \, ds = 0$.

Chapter 4 Hyperbolic Conservation Laws

4.1 Linear advection equation, characteristics method

In many applications, hyperbolic conservation laws arise when the diffusive terms are neglected from a more complete physical system, which is typically parabolic. A hyperbolic system usually better describes the wave phenomena, yet at the cost of losing regularity/smoothness of the solution. In fact, discontinuities are the major topic in the study of hyperbolic systems.

The simplest hyperbolic equation is a linear advection equation, which describes a wave propagating from left to right at a constant speed a,

$$u_t + au_x = 0. \tag{4.1}$$

It may be directly verified that the solution is

$$u(x,t) = u_0(x - at). \tag{4.2}$$

Alternatively, we consider the total derivative along a line $x = at + x_0$ as follows,

$$\left.\frac{du}{dt}\right|_{x=at+x_0} = \frac{\partial u}{\partial t} + \frac{\partial u}{\partial x} \cdot \frac{dx}{dt} = u_t + au_x = 0. \tag{4.3}$$

So u keeps unchanged along such a line. This line is called a characteristic line. The identification of such information and a path on which it propagates are key to the understanding of a general hyperbolic system.

In solving the linear advection equation, it is instrumental to draw the characteristic lines in the (x,t)-plane. As we shall see in later discussions, these lines describe the path of information propagation, and identify the value of $u(x,t)$ from the initial data for a nonlinear equation as well. See Fig. 4.1.

Furthermore, even when there is a jump in the initial profile, the formula $u(x,t) = u_0(x-at)$ still provides a function $u(x,t)$ in the half plane $(x,t) \in \mathbb{R} \times \mathbb{R}^+$. This function is not differentiable and does not satisfy the equation in the classical

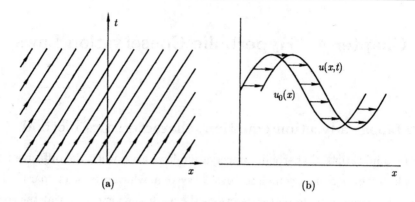

Figure 4.1 The linear advection equation: (a) characteristic lines; (b) initial data and the solution at a later time.

sense. Yet there are two ways to make a modification to the meaning of a solution. First, we may approximate the initial jump by a smooth curve in an arbitrarily short interval. The resulting function as the initial data then produces a classical solution. The classical solution, in the limit, gives a weak solution to the linear advection equation with the given initial data with a jump. Alternatively, we may regard the resulting function $u(x,t)$ with jump as a weak solution. Roughly speaking, it satisfies an integral form of the equation

$$\int_V u_t + a u_x \, dxdt = 0, \qquad (4.4)$$

over any controlling volume V in the (x,t) plane.

The characteristics method applies to general linear hyperbolic systems.

A first order linear PDE system in multiple space dimensions takes the form of

$$\sum_{i=1}^n \alpha_{ij} \frac{\partial u_j}{\partial x_i} + \beta_j u_j = 0, \quad j = 1, \cdots, m, \qquad (4.5)$$

where $\alpha_{ij}, \beta_j \in \mathbb{R}$. Superposition of solutions holds for a linear PDE system. If $u(x)$ and $\tilde{u}(x)$ are solutions, so is $\mu u(x) + \nu \tilde{u}(x)$, for any $\mu, \nu \in \mathbb{R}$.

A linear PDE system

$$\frac{\partial u}{\partial t} + A \frac{\partial u}{\partial x} = 0 \qquad (4.6)$$

Chapter 4 Hyperbolic Conservation Laws

is hyperbolic if A has distinct eigenvalues and a complete set of eigenvectors. In another word, there exists an invertible matrix P, such that A is diagonalized by

$$PAP^{-1} = \Lambda = \mathrm{diag}(\lambda_1, \cdots, \lambda_n). \tag{4.7}$$

We define a new set of variables $v = Pu$. Then it holds that

$$\frac{\partial v}{\partial t} + \Lambda \frac{\partial v}{\partial x} = 0. \tag{4.8}$$

We notice that the equations are now decoupled in the form of

$$\frac{\partial v_i}{\partial t} + \lambda_i \frac{\partial v_i}{\partial x} = 0. \tag{4.9}$$

In the same way as for the linear advection equation, each equation is solved by $v_i(x,t) = v_i(x - \lambda_i t, 0)$. Going back to the primary variables $u = P^{-1}v$, we obtain solution to the original linear system.

4.2 Nonlinear hyperbolic equations

We consider a fully nonlinear scalar equation with a source term

$$u_t + a(u; x, t) u_x = f(u; x, t). \tag{4.10}$$

It is straightforward to extend the characteristics method to this equation. As a matter of fact, along a curve

$$\Gamma : \frac{\mathrm{d}x}{\mathrm{d}t} = a(u; x, t), \quad x(0) = x_0, \tag{4.11}$$

it holds

$$\left.\frac{\mathrm{d}u}{\mathrm{d}t}\right|_\Gamma = \frac{\partial u}{\partial t} + \frac{\partial u}{\partial x} \cdot \frac{\mathrm{d}x}{\mathrm{d}t} = u_t + a u_x = f(u; x, t), \quad u(x_0, 0) = u_0(x_0). \tag{4.12}$$

Under certain fairly mild restrictions on $a(u; x, t)$, $f(u; x, t)$ and $u_0(x)$, the implicit function theorem provides a local existence result, i.e., for $t \leqslant \delta$, there exist $u(x, t)$ and $x(t)$ solving the ODE system (4.11)–(4.12).

The slope $a(u; x, t)$ of the characteristic curve identifies the propagation speed of information. It is usually bounded, leading to a finite propagation speed, which is a key feature in hyperbolic problems. In contrast, in a parabolic system such as the heat equation, any disturbance propagates instantly toward all positions. This explains why wave phenomena are not as clearly described by a parabolic system. We remark that finite propagation speed may also occur under certain special circumstances in a parabolic system.

Same as for the linear advection equation, we have an alternative, and maybe more important view to solve the initial value problem. That is, we seek for information paths over the half plane $(x,t) \in \mathbb{R} \times \mathbb{R}^+$. If the characteristic curves issued from the x-axis may cover this half plane, then the problem is solved.

In a general system of hyperbolic partial differential equations, this characteristics method may likely fail, e.g., in a gas dynamics system. Nevertheless, this approach is still a basic tool for constructing elementary waves there.

A system of quasilinear hyperbolic conservation laws reads

$$U_t + (F(U(x,t)))_x = 0, \qquad (4.13)$$

with $U(x,t) \in \mathbb{R}^n$, and $A(U) = \nabla F(U)$ is a diagonalizable $n \times n$ matrix. We call it hyperbolic because a perturbation around a constant state evolves in a similar way as the linear hyperbolic equations.

Here we list a few such systems.

First, we consider a traffic flow in a single-lane road. Let ρ, u be the density and velocity in a continuous sense, respectively. We further assume that a driver takes a speed according to how crowded the lane is. He takes a maximal speed $u_{\max}$ if there is no traffic, $\rho = 0$, and fully stops if the car density reaches a certain maximal value $\rho_{\max}$. The equation reads

$$\rho_t + (\rho u)_x = 0, \quad \text{with } u(\rho) = u_{\max}\left(1 - \frac{\rho}{\rho_{\max}}\right). \qquad (4.14)$$

Now consider a constant state $\bar{\rho}$, which can be arbitrarily chosen between 0 and $\rho_{\max}$. If we perform a perturbation with $\rho(x,t) = \tilde{\rho}(x,t) + \bar{\rho}$, then the small deviation evolves according to

$$\tilde{\rho}_t + u_{\max}\left(1 - \frac{2\bar{\rho}}{\rho_{\max}}\right)\tilde{\rho}_x = 0. \qquad (4.15)$$

This is a linear advection equation. When $2\bar{\rho} < \rho_{\max}$, perturbation propagates forward. The cars move forward and bring the perturbation forward. On the other hand, when $2\bar{\rho} > \rho_{\max}$, the traffic is quite full. Perturbation propagates backward, as actually drivers behind accelerate or brake according to behavior of cars in front of them.

As a second example, we illustrate the Buckley-Leverett equation for a two-phase flow. Let $u(x,t) \in [0,1]$ be the saturation of water in soil or porous media,

$$u_t + f(u)_x = 0, \quad f(u) = \frac{u^2}{u^2 + a(1-u)^2}. \tag{4.16}$$

It is found, around a constant saturation $\bar{u}$, a small perturbation propagates linearly at a speed

$$f'(\bar{u}) = \frac{2\bar{u}^2 \left[(1+a)\bar{u} + (1-a)\right]}{[\bar{u}^2 + a(1-\bar{u})^2]^2}. \tag{4.17}$$

Next, we describe the Euler equations for an isentropic polytropic gas flow in one space dimension. We take ρ, u, p the density, velocity and pressure, respectively. The dynamics is governed by

$$\begin{cases} \rho_t + (\rho u)_x = 0, \\ (\rho u)_t + (\rho u^2 + p)_x = 0. \end{cases} \tag{4.18}$$

Here the pressure $p(\rho) = k\rho^\gamma$ with $1 < \gamma < 3$. Again, perturbation around a constant state $(\bar{\rho}, \bar{u})$ evolves according to a linear hyperbolic system. Moreover, this property does not depend on the choice of primary variables.

We take the primary variable as $U = (\rho, m)^T$ with the momentum $m = \rho u$. The flux functions are $F(U) = \left(m, \dfrac{m^2}{\rho} + p\right)^T$.

We then rewrite the nonlinear conservation laws (4.18) into a primitive form

$$U_t + A(U)U_x = 0. \tag{4.19}$$

The coefficient matrix $A(U)$ reads

$$A(U) = \nabla_U F(U) = \begin{bmatrix} 0 & 1 \\ -\dfrac{m^2}{\rho^2} + p'(\rho) & \dfrac{2\rho u}{\rho} \end{bmatrix} = \begin{bmatrix} 0 & 1 \\ p' - u^2 & 2u \end{bmatrix}. \tag{4.20}$$

It is easy to find that the eigenvalues of A are $\lambda_\pm(\rho, m) = u \pm \sqrt{p'(\rho)}$. The eigenvectors are $(1, \lambda_+)^T$ and $(1, \lambda_-)^T$, respectively. Now the transform matrix reads

$$P = \begin{bmatrix} 1 & 1 \\ \lambda_+ & \lambda_- \end{bmatrix}^{-1}. \tag{4.21}$$

Due to nonlinearity, it does not follow the linear case to decouple the system. That is, because $\lambda_\pm$ depend on U, if we define $V = PU$, then we would not obtain $V_t + \Lambda V_x = 0$ with $\Lambda = \text{diag}(\lambda_+, \lambda_-)$ in general. Nevertheless, if we perturb a constant state $(\bar{\rho}, \bar{m})$, the small perturbation evolves according to

$$\tilde{V}_t + \Lambda \tilde{V}_x = 0, \quad \tilde{V} = \begin{bmatrix} 1 & 1 \\ \lambda_+(\bar{\rho}, \bar{m}) & \lambda_-(\bar{\rho}, \bar{m}) \end{bmatrix}^{-1} \begin{bmatrix} \tilde{\rho} \\ \tilde{m} \end{bmatrix}. \tag{4.22}$$

Now we take another set of primary variables $(\rho, u)^T$. The momentum equation may be rewritten as

$$u_t + u u_x + \frac{p'(\rho)}{\rho} \rho_x = 0. \tag{4.23}$$

The coefficient matrix is

$$B = \begin{bmatrix} u & \rho \\ \dfrac{p'(\rho)}{\rho} & u \end{bmatrix}. \tag{4.24}$$

The eigenvalues may be found again as $\lambda_\pm = u \pm \sqrt{p'(\rho)}$. The corresponding eigenvectors are $\left(1, \dfrac{\sqrt{p'}}{\rho}\right)^T$ and $\left(1, -\dfrac{\sqrt{p'}}{\rho}\right)^T$, respectively. Actually, $(d\rho, du) \parallel \left(1, \pm\dfrac{\sqrt{p'}}{\rho}\right)$ is equivalent to $(d\rho, dm) \parallel (1, u \pm \sqrt{p'})$.

The last example is the shallow water equations. Let (h, v) be the water elevation and velocity respectively. The governing equations read

$$\begin{cases} h_t + (hv)_x = 0, \\ v_t + (v^2/2 + gh)_x = 0. \end{cases} \tag{4.25}$$

Wave propagation in this system actually corresponds to that in the polytropic gas equations with $\gamma = 2$.

4.3 Discontinuities in inviscid Burgers' equation

In a quasilinear hyperbolic PDE system, which is nonlinear with respect to the unknown function and its derivatives, the most distinct phenomenon is the appearance of discontinuities, in particular, the shock waves. This is genuinely a nonlinear phenomenon, and does not arise in linear equations.

We illustrate this through a simple example of inviscid Burgers' equation

$$u_t + (u^2/2)_x = 0. \tag{4.26}$$

Under the assumption of smoothness, e.g., $u \in C^1$, it is equivalent to

$$u_t + u u_x = 0. \tag{4.27}$$

According to previous discussions, u maintains constant along a characteristic curve $\dfrac{\mathrm{d}x}{\mathrm{d}t} = u(x,t)$. This implies a constant slope and the characteristic curve is actually a straight line. Furthermore, the solution may be written as

$$u(x_0 + u_0(x_0)t, t) = u_0(x_0). \tag{4.28}$$

As we mentioned before, the solvability of this Cauchy problem is equivalent to: for each $(x,t) \in \mathbb{R} \times \mathbb{R}^+$, does there exist exactly one characteristic line passing through this point?

Unfortunately the answer is **NO** in general, regardless to the regularity of the initial profile. We illustrate with monotone profiles as follows.

First, we consider an increasing initial profile. We draw a triplet of figures in Fig. 4.2. The left subplot shows the characteristic lines. The central subplot shows the initial data and the wave profile in a later time t. The right subplot is a side-down curve of the flux function $u^2/2$ versus u. At a point (x,t) in the left subplot, the value of the function is $u(x,t)$ in the central subplot, and the characteristic line is parallel to the tangent line in the right subplot.

From the figures, we observe that for each $(x,t) \in \mathbb{R} \times \mathbb{R}^+$, the characteristic lines give a spreading picture. Consequently, there is one and only one characteristic line passing through any given point (x,t). Therefore a unique global solution is guaranteed. Moreover, when time evolves, the wave profile becomes flatter.

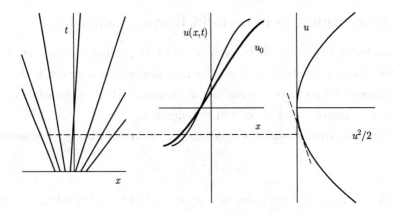

Figure 4.2   Inviscid Burgers' equation with increasing initial data.

Next, we consider a decreasing initial profile. We plot the triplet of figures in Fig. 4.3. The wave profile becomes sharper as time evolves. At a certain time, there must be a blow-up of the solution, that is, the spatial derivative $\partial u/\partial x$ becomes infinite at a certain point. If we keep following the characteristic lines, regardless of the fact that the original PDE becomes meaningless, then the wave profile will flip to the other side. For instance, we observe that the characteristic line through $(x^*, 0)$ goes across the point where $u_0 = 0$ at a certain time as shown in Fig. 4.3.

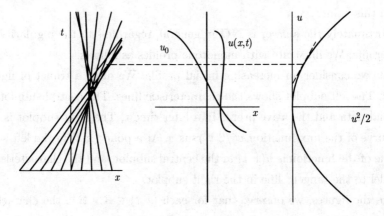

Figure 4.3   Inviscid Burgers' equation with decreasing initial data.

The above wave propagation may be explained by a queue of kids in a playground. If we put taller kids in the front, who usually run faster, then the queue becomes longer and longer as the shorter ones will fall behind. On the other hand, if we put the taller ones behind, then they will push the shorter ones and cause a crash.

The above discussions show that discontinuity is inevitable in general for a quasilinear hyperbolic equation. Classical solution is therefore not the whole story if we want to study a solution global in time. As a matter of fact, the hyperbolic equation does not hold at and after this blow-up time. We should refer to the physical problem, or the more complete system of convection-diffusion type. The discontinuity is then expanded in a finite yet small length scale, called shock layer, due to the diffusive terms.

In certain situations, for instance, in water waves, there might be some time when multiple-valued shape is acceptable. This multiple-valued shape does not hold for a long time, as the gravity will eventually push down the water drops. In most other cases, a discontinuity forms and moves at certain velocity. Along this discontinuity $x = x^*(t)$, the equation does not hold in the classical sense ($u_x(x^*(t), t) = \infty$). Away from that, the hyperbolic equation still holds in the classical sense.

In the following, we shall resolve the difficulty in a hyperbolic equation manner, following the pioneering work of Riemann in 1860's. Instead of avoiding discontinuities, we consider an equation with discontinuous initial data. It is of great importance that the Riemann problem serves as a building block for solving such a system, making use of the finite wave propagating speed.

4.4 Elementary waves in inviscid Burgers' equation

For inviscid Burgers' equation

$$u_t + uu_x = 0, \qquad (4.29)$$

we consider a Riemann problem, namely, a discontinuous Cauchy problem with piecewise constant initial data

$$u_0(x) = \begin{cases} u_-, & x < 0, \\ u_+, & x > 0. \end{cases} \qquad (4.30)$$

Not expecting a solution in the classical sense, we discuss piecewise smooth solution instead. This gives rise to two types of solutions, namely, the shock waves and the centered rarefaction waves.

There are different ways to introduce a shock wave. Here we present a vanishing viscosity approach. We speculate that a shock moves at a certain speed. It is a limit case of Burgers' equation

$$u_t + uu_x = \varepsilon u_{xx}. \tag{4.31}$$

A simplest viscous profile of the discontinuity corresponds to a traveling wave, with a transitional shock layer of width related to ε. If one starts with initial profile different from the traveling wave, the solution for Burgers' equation is a perturbation of the traveling wave. The perturbation propagates out of the shock layer asymptotically, possibly leading to a shift of the shock layer.

In the following, we seek for a heteroclinic orbit that connects u_- with u_+ in the traveling wave equation with $\xi = x - ct$,

$$-cu' + uu' = \varepsilon u''. \tag{4.32}$$

This can be integrated once to give

$$\varepsilon u' = -cu + u^2/2 + C_0. \tag{4.33}$$

Because u_- and u_+ are critical points, we have

$$-cu_- + u_-^2/2 + C_0 = 0, \quad -cu_+ + u_+^2/2 + C_0 = 0. \tag{4.34}$$

From these, we obtain

$$c = \frac{u_- + u_+}{2}, \quad C_0 = \frac{u_- u_+}{2}. \tag{4.35}$$

Furthermore, because $-cu + u^2/2 + C_0$ is convex, the source term is negative for u between u_- and u_+. Therefore, a heteroclinic orbit exists if and only if $u_- > u_+$. In another word, there is a shock if and only if $u_- > u_+$, at the propagation speed $c = (u_- + u_+)/2$.

The above results have two equivalent expressions. One is algebraic flavored. We introduce a notation for jump of an entity q as $[q] = q_+ - q_-$, where $q_\pm$ are the values across the discontinuity. Naturally, the shock strength is represented by $[u]$. The previous result may be rephrased as follows,

$$-c\,[u] + [u^2]/2 = 0. \tag{4.36}$$

This is called the Rankine-Hugoniot relation. Meanwhile, the entropy condition reads $u_- > c > u_+$, as the shock speed is between the characteristic speeds across the shock front.

An alternative expression is geometrical flavored. The shock speed is the slope of the secant line between $(u_-, u_-^2/2)$ and $(u_+, u_+^2/2)$. Moreover, the entropy condition $u_- > u_+$ is equivalent to that the characteristic lines from both sides point towards the shock front. Drawing the characteristic lines at both sides of the shock front, we observe a 2-in-0-out picture if viewing from the shock front. This is the Lax entropy condition. See Fig. 4.4.

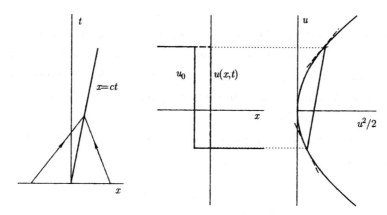

Figure 4.4 Shock in inviscid Burgers' equation.

We conclude that the shock solution for the Riemann problem reads

$$u(x,t) = \begin{cases} u_-, & x < ct, \\ u_+, & x > ct. \end{cases} \tag{4.37}$$

Another type of elementary wave is a centered rarefaction wave. If $u_- < u_+$, there is no shock wave as the entropy condition is violated. Instead, since the characteristic lines issued from $x < 0$ only cover the domain $x < u_- t$; and those issued from $x > 0$ only cover the domain $x > u_+ t$, there is a region $u_- t < x < u_+ t$ where there is no definition for the solution.

There are two different ways to find the exact form of the rarefaction.

The first way is to find a self-similar solution $u(x,t) = u(\xi)$ in terms of a self-similar variable $\xi = x/t$. Such solution exists as the equation and the initial data remains unchanged under the transform $(u; x, t) \longrightarrow (u; \alpha x, \alpha t)$, where α is an arbitrary number. We remark that the shock wave solution (4.37) is also a self-similar solution.

The equation is then reduced to

$$-\frac{xu'}{t^2} + \frac{uu'}{t} = 0. \tag{4.38}$$

Or, equivalently,

$$(u - \xi)u' = 0. \tag{4.39}$$

There are two possibilities. One is $u' = 0$ thus u is constant. It is of no interest here for solving a discontinuous problem. The second choice is $u = \xi$. This gives a fan. On each line $x = \xi t$, we have $u = \xi$. We remark that at the origin, $u(0,0)$ is not defined. A comprehensive understanding relies on the definition of a weak solution, which we omit here.

The solution to the Riemann problem is defined by piecing together the constant states and a sector of the rarefaction fan,

$$u(x,t) = \begin{cases} u_-, & \text{if } x < u_- t, \\ x/t, & \text{if } u_- t < x < u_+ t, \\ u_+, & \text{if } x > u_+ t. \end{cases} \tag{4.40}$$

This is discontinuous not only at the origin, but also at the borders of the centered rarefaction fan $x = u_\pm t$, where the partial derivatives do not exist. See Fig. 4.5.

The other approach is to imagine a small perturbation of the initial data, e.g.,

$$u(x,0) = \begin{cases} u_-, & \text{if } x < u_- \varepsilon, \\ x/\varepsilon, & \text{if } u_- \varepsilon < x < u_+ \varepsilon, \\ u_+, & \text{if } x > u_+ \varepsilon. \end{cases} \tag{4.41}$$

Chapter 4 Hyperbolic Conservation Laws

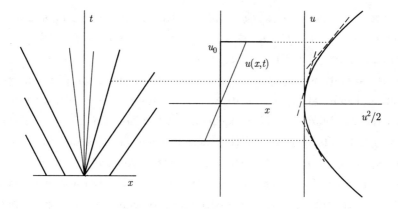

Figure 4.5 Rarefaction wave in inviscid Burgers' equation.

The characteristics method then applies to produce a solution

$$u(x,t) = \begin{cases} u_-, & \text{if } x < u_-(t+\varepsilon), \\ x/(t+\varepsilon), & \text{if } u_-(t+\varepsilon) < x < u_+(t+\varepsilon), \\ u_+, & \text{if } x > u_+(t+\varepsilon). \end{cases} \qquad (4.42)$$

Taking the limit $\varepsilon \to 0$, we obtain the same solution (4.40).

We remark that the smoothed initial data may be viewed as the viscous picture for the complete Burgers' equation. In fact, with the discontinuous Riemann data, the viscosity yields a smooth profile immediately afterwards. It is worth mentioning that with different monotone perturbations for the initial data, we obtain the same solution in the limit $\varepsilon \to 0$. This justifies partially our solution with centered rarefaction wave.

4.5 Wave interactions in inviscid Burgers' equation

With the elementary waves, we are ready to investigate more general initial data. In particular, we consider wave interaction problems with a piecewise constant initial data

$$u(x,0) = \begin{cases} u_-, & \text{if } x < a, \\ u_m, & \text{if } a < x < b, \\ u_+, & \text{if } x > b. \end{cases} \qquad (4.43)$$

Six cases may arise, according to the relative position of u_-, u_m and u_+. They are: (A) $u_- > u_m > u_+$; (B) $u_- > u_+ > u_m$; (C) $u_m > u_- > u_+$; (D) $u_m > u_+ > u_-$; (E) $u_+ > u_- > u_m$; (F) $u_+ > u_m > u_-$.

Let us illustrate with (A) $u_- > u_m > u_+$. Due to finite propagation speed of information, locally at $(x,t) = (a,0)$, we solve a Riemann problem with piecewise constant data u_- and u_m. There forms a shock denoted as S_1 with speed $s_1 = (u_- + u_m)/2$. Similarly, around $(b,0)$, there forms a second shock S_2 propagating at speed $s_2 = (u_m + u_+)/2$. At a small time t, the two shock fronts are located at $x = a + s_1 t$, and $x = b + s_2 t$, respectively. At any time t before these two wave fronts meet, locally we still have the same Riemann problems due to finite propagation speed. So the two shock profile maintains until the wave fronts meet, or, when the shocks S_1 and S_2 collide.

In summary, the solution is piecewise constant before collision,

$$u(x,t) = \begin{cases} u_-, & \text{if } x < a + s_1 t, \\ u_m, & \text{if } a + s_1 t < x < b + s_2 t, \\ u_+, & \text{if } x > b + s_2 t. \end{cases} \qquad (4.44)$$

Since $s_1 > s_2$, the first shock S_1 catches up with the second one at time

$$t^* = \frac{b-a}{s_1 - s_2} = \frac{2(b-a)}{u_- - u_+}, \qquad (4.45)$$

and position

$$x^* = \frac{s_2 a - s_1 b}{s_2 - s_1}. \qquad (4.46)$$

At the collision time, we solve a new Riemann problem

$$u(x, t^*) = \begin{cases} u_-, & \text{if } x < x^*, \\ u_+, & \text{if } x > x^*. \end{cases} \qquad (4.47)$$

Because $u_- > u_+$, there appears a new shock S^* at $x = x^* + s^*(t - t^*)$, with shock speed $s^* = (u_- + u_+)/2$. We observe that in the $(u, f(u))$-plane subplot of Fig. 4.6, there is a triangle with vertices corresponding to these three shocks: S_1, S_2 and S^*.

This type of wave interaction is denoted formally as $S + S \longrightarrow S$. The solution is summarized as follows.

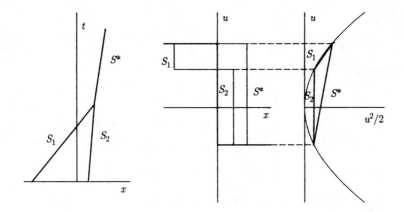

Figure 4.6 Shock collision in inviscid Burgers' equation.

When $0 < t < \dfrac{2(b-a)}{u_- - u_+}$, we have

$$u(x,t) = \begin{cases} u_-, & \text{if } x < a + \dfrac{u_- + u_m}{2}t, \\ u_m, & \text{if } a + \dfrac{u_- + u_m}{2}t < x < b + \dfrac{u_m + u_+}{2}t, \\ u_+, & \text{if } x > b + \dfrac{u_m + u_+}{2}t. \end{cases} \qquad (4.48)$$

When $t > \dfrac{2(b-a)}{u_- - u_+}$, we have

$$u(x,t) = \begin{cases} u_-, & \text{if } x < \dfrac{u_- + u_+}{2}t + \dfrac{a(u_- - u_m) + b(u_m - u_+)}{u_- - u_+}, \\ u_+, & \text{if } x > \dfrac{u_- + u_+}{2}t + \dfrac{a(u_- - u_m) + b(u_m - u_+)}{u_- - u_+}. \end{cases} \qquad (4.49)$$

Now we consider (E) $u_+ > u_- > u_m$. Initially, there are a shock S_1 connecting u_- and u_m, and a centered rarefaction wave connecting u_m and u_+. Because the shock speed is faster than the first characteristic line in the rarefaction wave, i.e.,

$$s_- = \frac{u_- + u_m}{2} > u_m, \qquad (4.50)$$

the shock will hit the rarefaction fan at time

$$t^* = \frac{2(b-a)}{u_- - u_m}, \qquad (4.51)$$

and position
$$x^* = a + s_- t^* = a + \frac{(b-a)(u_- + u_m)}{u_- - u_m}. \tag{4.52}$$

The solution for $t < t^*$ is therefore
$$u(x,t) = \begin{cases} u_-, & \text{if } x < a + s_- t, \\ u_m, & \text{if } a + s_- t < x < b + u_m t, \\ \dfrac{x-b}{t}, & \text{if } b + u_m t < x < b + u_+ t, \\ u_+, & \text{if } x > b + u_+ t. \end{cases} \tag{4.53}$$

Afterwards, the shock 'bites' gradually the rarefaction. In this process, the shock strength $[u]$ weakens. To get the solution, we denote the shock front position as $x_s(t)$, and the value on the right side of the shock front as $u_s(t) = \lim_{x \to x_s(t)+} u(x,t)$. From the Rankine-Hugoniot relation and the rarefaction wave solution, we have
$$\begin{cases} \dfrac{dx_s(t)}{dt} = \dfrac{u_- + u_s(t)}{2}, \\ u_s(t) = \dfrac{x_s(t) - b}{t}. \end{cases} \tag{4.54}$$

Solving the ordinary differential equation, we obtain
$$x_s(t) = C\sqrt{t} + u_- t + b. \tag{4.55}$$

With the condition $x_s(t^*) = x^*$, we find that
$$C = (u_m - u_-)\sqrt{t^*} = -\sqrt{2(b-a)(u_- - u_m)}. \tag{4.56}$$

Therefore, the shock front locates at
$$x_s(t) = (u_m - u_-)\sqrt{tt^*} + u_- t + b. \tag{4.57}$$

Moreover, we have
$$u_s(t) = u_- + (u_m - u_-)\sqrt{\frac{t^*}{t}}. \tag{4.58}$$

Noticing that $u_s(t) > u_-$ for all time, and it approaches to equality asymptotically, we conclude that the shock decays along with time, on the order of $[u] \sim t^{-1/2}$. See Fig. 4.7.

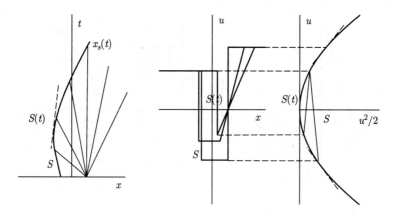

Figure 4.7 Shock partially overtakes rarefaction wave in inviscid Burgers' equation.

The solution reads as follows, for $t > t^*$,

$$u(x,t) = \begin{cases} u_-, & \text{if } x < x_s(t), \\ \dfrac{x-b}{t}, & \text{if } x_s(t) < x < b + u_+ t, \\ u_+, & \text{if } x > b + u_+ t. \end{cases} \quad (4.59)$$

Next, we consider (F) $u_- < u_m < u_+$. We have initially two rarefaction waves R_1 and R_2. Since the last characteristic line in R_1 takes the same slope as the first characteristic line in R_2, there is no interaction. That is, the solution is simply a combination of these two rarefaction wave solutions. See Fig. 4.8. The solution reads

$$u(x,t) = \begin{cases} u_-, & \text{if } x < a + u_- t, \\ \dfrac{x-a}{t}, & \text{if } a + u_- t < x < a + u_m t, \\ u_m, & \text{if } a + u_m t < x < b + u_m t, \\ \dfrac{x-b}{t}, & \text{if } b + u_m t < x < b + u_+ t, \\ u_+, & \text{if } x > b + u_+ t. \end{cases} \quad (4.60)$$

We leave discussions for the rest three cases to the readers. Here we only present the conclusions.

For (B) $u_- > u_+ > u_m$, a shock overtakes a rarefaction completely. When t is big enough, there is only a shock wave connecting u_- and u_+. The solution is as follows.

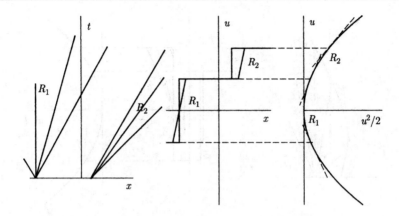

Figure 4.8 Two rarefaction waves with no interaction in inviscid Burgers' equation.

When $0 < t < \dfrac{2(b-a)}{u_- - u_m}$, we have

$$u(x,t) = \begin{cases} u_-, & \text{if } x < a + \dfrac{u_- + u_m}{2}t, \\ u_m, & \text{if } a + \dfrac{u_- + u_m}{2}t < x < b + u_m t, \\ \dfrac{x-b}{t}, & \text{if } b + u_m t < x < b + u_+ t, \\ u_+, & \text{if } x > b + u_+ t. \end{cases} \qquad (4.61)$$

When $\dfrac{2(b-a)}{u_- - u_m} < t < \dfrac{2(b-a)(u_- - u_m)}{(u_- - u_+)^2}$, we have

$$u(x,t) = \begin{cases} u_-, & \text{if } x < u_- t - \sqrt{2(b-a)(u_- - u_m)t} + b, \\ \dfrac{x-b}{t}, & \text{if } u_- t - \sqrt{2(b-a)(u_- - u_m)t} + b < x < b + u_+ t, \\ u_+, & \text{if } x > b + u_+ t. \end{cases} \qquad (4.62)$$

When $t > \dfrac{2(b-a)(u_- - u_m)}{(u_- - u_+)^2}$, we have

$$u(x,t) = \begin{cases} u_-, & \text{if } x < \dfrac{u_- + u_+}{2}t + \dfrac{a(u_- - u_m) + b(u_m - u_+)}{u_- - u_+}, \\ u_+, & \text{if } x > \dfrac{u_- + u_+}{2}t + \dfrac{a(u_- - u_m) + b(u_m - u_+)}{u_- - u_+}. \end{cases} \qquad (4.63)$$

For (C) $u_m > u_- > u_+$, a shock overtakes a rarefaction completely. It is essentially the same as (B). The solution is as follows.

When $0 < t < \dfrac{2(b-a)}{u_m - u_+}$, we have

$$u(x,t) = \begin{cases} u_-, & \text{if } x < a + u_- t, \\ \dfrac{x-a}{t}, & \text{if } a + u_- t < x < a + u_m t, \\ u_m, & \text{if } a + u_m t < x < b + \dfrac{u_m + u_+}{2} t, \\ u_+, & \text{if } x > b + \dfrac{u_m + u_+}{2} t. \end{cases} \qquad (4.64)$$

When $\dfrac{2(b-a)}{u_m - u_+} < t < \dfrac{2(b-a)(u_m - u_+)}{(u_- - u_+)^2}$, we have

$$u(x,t) = \begin{cases} u_-, & \text{if } x < a + u_- t, \\ \dfrac{x-a}{t}, & \text{if } a + u_- t < x < u_+ t + \sqrt{2(b-a)(u_m - u_+)t} + a, \\ u_+, & \text{if } x > u_+ t + \sqrt{2(b-a)(u_m - u_+)t} + a. \end{cases} \qquad (4.65)$$

When $t > \dfrac{2(b-a)(u_m - u_+)}{(u_- - u_+)^2}$, we have

$$u(x,t) = \begin{cases} u_-, & \text{if } x < \dfrac{u_- + u_+}{2} t + \dfrac{a(u_- - u_m) + b(u_m - u_+)}{u_- - u_+}, \\ u_+, & \text{if } x > \dfrac{u_- + u_+}{2} t + \dfrac{a(u_- - u_m) + b(u_m - u_+)}{u_- - u_+}. \end{cases} \qquad (4.66)$$

For (D) $u_m > u_+ > u_-$, the initial wave structure contains a rarefaction and a shock. The shock strength decays at the rate $t^{-1/2}$, and the rarefaction persists for all the time. It is essentially the same as (E).

When $0 < t < \dfrac{2(b-a)}{u_m - u_+}$, the solution is

$$u(x,t) = \begin{cases} u_-, & \text{if } x < a + u_- t, \\ \dfrac{x-a}{t}, & \text{if } a + u_- t < x < a + u_m t, \\ u_m, & \text{if } a + u_m t < x < b + \dfrac{u_m + u_+}{2} t, \\ u_+, & \text{if } x > b + \dfrac{u_m + u_+}{2} t. \end{cases} \qquad (4.67)$$

When $t > \dfrac{2(b-a)}{u_m - u_+}$, the solution reads

$$u(x,t) = \begin{cases} u_-, & \text{if } x < a + u_-t, \\ \dfrac{x-a}{t}, & \text{if } a + u_-t < x < a + u_+t + \sqrt{2(b-a)(u_m - u_+)t}, \\ u_+, & \text{if } x > a + u_+t + \sqrt{2(b-a)(u_m - u_+)t}. \end{cases} \quad (4.68)$$

We remark that in all above cases, the asymptotic behavior of the system is always dictated by the Riemann problem with (u_-, u_+).

4.6 Elementary waves in a polytropic gas

For a polytropic gas described in (4.18), it turns out that the so-called Lagrangian formulation suits better for mathematical analysis. In fact, under the assumption that $\rho u|_{x=x_0} = 0$, we define

$$\begin{cases} \tau = t, \\ \xi = \displaystyle\int_{x_0}^{x} \rho(y, t) dy. \end{cases} \quad (4.69)$$

By virtue of the equation for mass conservation, ξ actually identifies the 'particle'. In fact, in one space dimension, no particle can go across a 'neighbor' particle. Therefore, the total mass at its left side is an index for this particle. For this transform, we have,

$$\begin{cases} \dfrac{\partial}{\partial t} = \dfrac{\partial}{\partial \tau} + \displaystyle\int_{x_0}^{x} \rho_t(y, t) dy \dfrac{\partial}{\partial \xi} = \dfrac{\partial}{\partial \tau} - \rho u \dfrac{\partial}{\partial \xi}, \\ \dfrac{\partial}{\partial x} = \rho \dfrac{\partial}{\partial \xi}. \end{cases} \quad (4.70)$$

The mass conservation equation becomes

$$\rho_t + m_x = \rho_\tau - m\rho_\xi + \rho m_\xi = -\rho^2\left[(1/\rho)_\tau - (m/\rho)_\xi\right] = 0. \quad (4.71)$$

Meanwhile, noticing that the equation for momentum conservation leads to

$$u_t + u u_x + p_x/\rho = 0, \quad (4.72)$$

we have

$$u_t + u u_x + p_x/\rho = u_\tau - m u_\xi + u \rho u_\xi + p_\xi = u_\tau + p_\xi = 0. \quad (4.73)$$

So, the p-system in the Lagrangian coordinates reads

$$\begin{cases} v_\tau - u_\xi = 0, \\ u_\tau + p_\xi = 0. \end{cases} \quad (4.74)$$

Here we define specific volume $v = 1/\rho$. Again we consider linearization around a ground state (v_0, u_0) as follows,

$$\begin{bmatrix} v \\ u \end{bmatrix}_\tau + \begin{bmatrix} 0 & -1 \\ p'(v_0) & 0 \end{bmatrix} \begin{bmatrix} v \\ u \end{bmatrix}_\xi = h.o.t. \quad (4.75)$$

We remark that the derivative for pressure is taken with respect to v. That is, $p'(v_0) = -\gamma v_0^{-(\gamma+1)}$. The two distinct eigenvalues are

$$\lambda_\pm = \pm\sqrt{-p'(v_0)} = \pm\lambda; \quad \lambda(v_0) = \sqrt{\gamma}v_0^{-(\gamma+1)/2}. \quad (4.76)$$

For hyperbolic conservation laws, we may locally follow the strategy in linear case to diagonalize the system as follows,

$$U_t + (F(U))_x = 0 \Rightarrow V_t + \Lambda V_x = h.o.t. \quad (4.77)$$

Here Λ is a diagonal matrix. But the high order terms can not be discarded, particularly when we investigate shocks, which is not a small perturbation from any ground state.

It is striking that we may find Riemann variables in this two equation system. In fact, we define

$$\begin{cases} r = u - \int \lambda dv = u + \dfrac{2\sqrt{\gamma}}{\gamma - 1} v^{(1-\gamma)/2}, \\ s = u + \int \lambda dv = u - \dfrac{2\sqrt{\gamma}}{\gamma - 1} v^{(1-\gamma)/2}. \end{cases} \quad (4.78)$$

Conversely, we may recover (v, u) from

$$\begin{cases} u = (r + s)/2, \\ v = \left[(\gamma - 1)(r - s)/4\sqrt{\gamma}\right]^{(\gamma-1)/2}. \end{cases} \quad (4.79)$$

By a straight forward calculation, we find

$$r_\tau + \lambda r_\xi = (u_\tau - \lambda v_\tau) + \lambda(u_\xi - \lambda v_\xi) = u_\tau + p'(v)v_\xi = 0. \quad (4.80)$$

Similarly, we have

$$s_\tau - \lambda s_\xi = (u_\tau + \lambda v_\tau) - \lambda(u_\xi + \lambda v_\xi) = u_\tau + p'(v)v_\xi = 0. \tag{4.81}$$

Therefore, we have a right-going characteristic curve

$$\Gamma_1 : \quad \frac{d\xi}{d\tau} = \lambda, \tag{4.82}$$

along which r keeps constant. Along a left-going characteristic curve

$$\Gamma_2 : \quad \frac{d\xi}{d\tau} = -\lambda, \tag{4.83}$$

s keeps constant.

The coupling of (r,s) hides in the expression of v and therefore λ. We notice that λ depends on v, which may be expressed in terms of $r - s$. Therefore, along a right-going characteristic curve, s varies while r remains unchanged. Hence v changes, yielding a changing λ. The characteristic curve is not a straight line in general.

In physical problems, typically we have certain upper-bound for density, this gives an upper-bound $\lambda_{\max}$ for λ. That is, we have finite speed of propagation in the p-system. So we also have the notions of domain of dependence and range of influence. In fact, at each point (ξ_0, τ_0), the physical quantities depend at most $D = [\xi_0 - \lambda_{\max}\tau_0, \xi_0 - \lambda_{\max}\tau_0]$. Meanwhile, the quantities here influence at most the domain $R = \{ (\xi, \tau) | \ |\xi - \xi_0| \leqslant \lambda_{\max}|\tau - \tau_0|\}$. See Fig. 4.9.

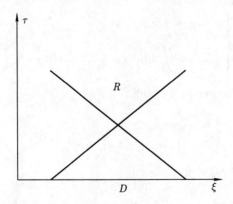

Figure 4.9 Domain of dependence and range of influence.

Now we describe elementary waves in the p-system.

For a shock wave, we may consider a viscous fluid

$$\begin{cases} v_\tau - u_\xi = 0, \\ u_\tau + p_\xi = \nu u_{\xi\xi}. \end{cases} \quad (4.84)$$

A traveling wave with speed s solves a traveling wave equation with $\eta = \xi - s\tau$,

$$\begin{cases} -sv' - u' = 0, \\ -su' + p' = \nu u''. \end{cases} \quad (4.85)$$

They can be readily integrated once to give

$$\begin{cases} -sv - u = C_1, \\ -su + p + C_2 = \nu u'. \end{cases} \quad (4.86)$$

The Riemann data $(u_\pm, v_\pm)$ must be critical points, and we obtain the Rankine-Hugoniot relations (with $[q] = q_+ - q_-$)

$$\begin{cases} -s[v] - [u] = 0, \\ -s[u] + [p] = 0. \end{cases} \quad (4.87)$$

Eliminating the term $[u]$, we obtain

$$s_\pm = \pm\sqrt{-\frac{[p]}{[v]}}. \quad (4.88)$$

We remark that in the limit of continuous case $[v] \to 0$, this gives the eigenvalue $\lambda_\pm$.

Now we need to determine a condition for existence of such a heteroclinic orbit. We may easily obtain

$$\nu s v' = -s^2 v - p + C_3 = \frac{[p]}{[v]} v - p + C_3. \quad (4.89)$$

Because of the convexity of $p(v)$, we know that for v between v_- and v_+, the right hand side is positive.

We first consider s_+. Then a heteroclinic orbit exists if and only if $v_- < v_+$. This inequality in fact is equivalent to $\lambda_+(v_-) > \lambda_+(v_+)$. Therefore we obtain a right-going (forward) shock wave.

Across $S_+ : \xi = s_+\tau$, we have

$$(v, u) = \begin{cases} (v_-, u_-), & \text{if } \xi < s_+\tau, \\ (v_+, u_+), & \text{if } \xi > s_+\tau, \end{cases} \quad (4.90)$$

with

$$v_- < v_+, \quad s_+ = \sqrt{-[p]/[v]}, \quad [u] = -\sqrt{-[p][v]}. \quad (4.91)$$

Please refer to Fig. 4.10 for comprehension of the forward shock. In the subplot (b), the solid line identifies right end-states that can be connected to (v_-, u_-) by a forward shock, while the dashed line identifies left end-states that can be connected to (v_+, u_+) by a forward shock.

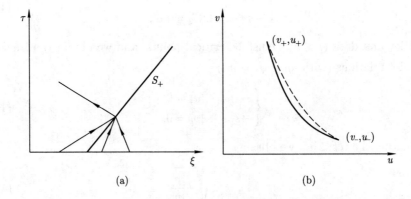

Figure 4.10 Forward shock wave in polytropic gas: (a) physical space; (b) state space.

For the case of s_-, the condition for the existence of a heteroclinic orbit is $v_- > v_+$, which is equivalent to say $\lambda_-(v_-) > \lambda_-(v_+)$. A left-going (backward) shock wave solution is

$$(v, u) = \begin{cases} (v_-, u_-), & \text{if } \xi < s_-\tau, \\ (v_+, u_+), & \text{if } \xi > s_-\tau, \end{cases} \quad (4.92)$$

with

$$v_- > v_+, \quad s_- = -\sqrt{-[p]/[v]}, \quad [u] = -\sqrt{-[p][v]}. \quad (4.93)$$

Please refer to Fig. 4.11 for comprehension of the backward shock. In the subplot (b), the solid line identifies right end-states that can be connected to (v_-, u_-)

by a backward shock, while the dashed line identifies left end-states that can be connected to (v_+, u_+) by a backward shock.

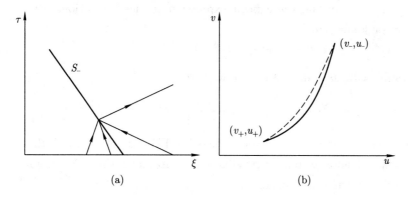

Figure 4.11   Backward shock wave in polytropic gas: (a) physical space; (b) state space.

In these two cases, the entropy conditions may be uniformly stated in terms of the Lax entropy condition, namely, there are "3-in-1-out" characteristic lines.

Now we look for rarefaction waves. Taking the self-similar variable $\eta = \xi/\tau$, we have
$$\begin{cases} -\eta v^{\scriptscriptstyle\centerdot} - u^{\scriptscriptstyle\centerdot} = 0, \\ -\eta u^{\scriptscriptstyle\centerdot} + p'(v) v^{\scriptscriptstyle\centerdot} = 0. \end{cases} \tag{4.94}$$

This can be viewed as a linear system for variables $(v^{\scriptscriptstyle\centerdot}, u^{\scriptscriptstyle\centerdot})$. Either $v^{\scriptscriptstyle\centerdot} = u^{\scriptscriptstyle\centerdot} = 0$, which is discarded, or the coefficient matrix is singular. Therefore,

$$\det \begin{bmatrix} -\eta & -1 \\ p'(v) & -\eta \end{bmatrix} = \eta^2 + p'(v) = 0. \tag{4.95}$$

This means $\eta = \pm \lambda(v)$. For the case $\eta = \lambda(v)$, we have a forward rarefaction wave R_+, which satisfies
$$-\lambda v^{\scriptscriptstyle\centerdot} - u^{\scriptscriptstyle\centerdot} = 0. \tag{4.96}$$

This leads to, in the (v, u) plane,
$$s^{\scriptscriptstyle\centerdot} = \lambda v^{\scriptscriptstyle\centerdot} + u^{\scriptscriptstyle\centerdot} = 0. \tag{4.97}$$

So s keeps constant along this type of rarefaction wave.

Accordingly, in the physical plane ((ξ, τ) plane), from left to right, η increases. Therefore, λ increases, and v decreases. So, r and u increase.

In fact, we may find the explicit expression of such a solution. Since $\lambda(v) = \eta = \xi/\tau$, we know that

$$v(\eta) = \left(\eta^2/\gamma\right)^{-1/(\gamma+1)}. \tag{4.98}$$

Meanwhile, with $\mathrm{d}s = \mathrm{d}u + \lambda \mathrm{d}v = 0$, we find

$$u = s_0 + \frac{2\sqrt{\gamma}}{\gamma - 1} v^{(1-\gamma)/2} = u_0 + \frac{2\sqrt{\gamma}}{\gamma - 1}\left(v^{(1-\gamma)/2} - v_0^{(1-\gamma)/2}\right). \tag{4.99}$$

The forward rarefaction wave is depicted in Fig. 4.12. We remark that we only have one curve here to connect possible states. Actually, this curve is governed by a constant s in the (v, u) plane.

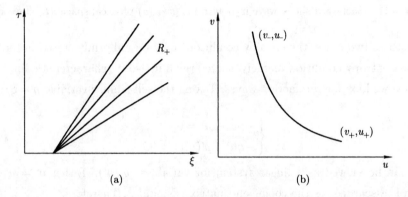

Figure 4.12 Forward rarefaction wave in polytropic gas: (a) physical space; (b) state space.

For the other case $\eta = -\lambda(v)$, we have a backward rarefaction wave R_-, which satisfies

$$\lambda v^{\prime} - u^{\prime} = 0. \tag{4.100}$$

This leads to, in the (v, u) plane,

$$r^{\prime} = -\lambda v^{\prime} + u^{\prime} = 0. \tag{4.101}$$

So r keeps constant along this type of rarefaction wave. By a similar argument we may find that s, v and u increases.

Again we may find the explicit expression of such a solution. Now $\lambda(v) = -\eta = -\xi/\tau$, we know that
$$v(\eta) = \left(\eta^2/\gamma\right)^{-1/(\gamma+1)}. \tag{4.102}$$
Meanwhile, with $dr = du - \lambda dv = 0$, we find
$$u = r_0 - \frac{2\sqrt{\gamma}}{\gamma - 1} v^{(1-\gamma)/2} = u_0 - \frac{2\sqrt{\gamma}}{\gamma - 1} \left(v^{(1-\gamma)/2} - v_0^{(1-\gamma)/2}\right). \tag{4.103}$$
Please see Fig. 4.13.

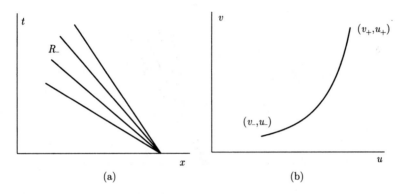

Figure 4.13 Backward rarefaction wave in polytropic gas: (a) physical space; (b) state space.

4.7 Riemann problem in a polytropic gas

For a polytropic gas in Lagrangian coordinates
$$\begin{cases} v_\tau - u_\xi = 0, \\ u_\tau + p_\xi = 0, \end{cases} \tag{4.104}$$
we consider a Riemann problem with initial data
$$(v(x,0), u(x,0)) = \begin{cases} (v_-, u_-), & \text{if } x < 0, \\ (v_+, u_+), & \text{if } x > 0. \end{cases} \tag{4.105}$$
To solve it, we need to find elementary waves connecting these two states. Since both the equation and the initial data possess invariance under the transform

$\{\xi \to \alpha\xi, \tau \to \alpha\tau\}$, so the solution must depend only on $\eta = \xi/\tau$. For the left quadrant $\eta < 0$, if there is a shock at $\eta = s_- < 0$, due to the entropy condition, there cannot be a backward rarefaction wave. For the same reason, there is at most one type of forward wave on the right quadrant. Consequently, to solve a Riemann problem, we only need to find forward and backward waves. It amounts to find an intermediate state (v^*, u^*). A backward wave connects (v_-, u_-) with this state, and a forward wave connects it with (v_+, u_+). See Fig. 4.14 for illustration.

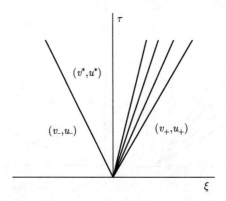

Figure 4.14 Riemann solution in the physical space.

Now we consider a backward wave that takes (v_-, u_-) as left end state, possibly a shock as well as a rarefaction wave. Then we have

$$u^* - u_- \equiv f_-(v^*; v_-) = \begin{cases} -\sqrt{-(p(v^*) - p(v_-))(v^* - v_-)}, & \text{if } v_- > v^*, \\ -\dfrac{2\sqrt{\gamma}}{\gamma - 1}\left(v^{*(1-\gamma)/2} - v_-^{(1-\gamma)/2}\right), & \text{if } v_- < v^*. \end{cases} \quad (4.106)$$

Similarly, we consider a forward wave that takes (v_+, u_+) as left end state. See Fig. 4.15. Then we have

$$u^* - u_+ \equiv f_+(v^*; v_+) = \begin{cases} \sqrt{-(p(v_+) - p(v^*))(v_+ - v^*)}, & \text{if } v^* < v_+, \\ -\dfrac{2\sqrt{\gamma}}{\gamma - 1}\left(v_+^{(1-\gamma)/2} - v^{*(1-\gamma)/2}\right), & \text{if } v^* > v_+. \end{cases} \quad (4.107)$$

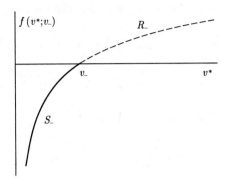

Figure 4.15 Backward wave in the phase plane.

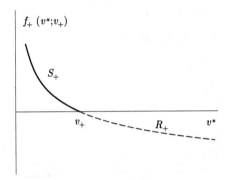

Figure 4.16 Forward wave in the phase plane.

Given the Riemann data, we may eliminate u^* from the above equations, and obtain an equation solely of v^*,

$$u_+ - u_- = f_-(v^*; v_-) - f_+(v^*; v_+); \qquad (4.108)$$

This is equivalent to find an intersection point of the curves $f_-(v^*; v_-)$ and $f_+(v^*; v_+)$, up to a translation of $(u_+ - u_-)$. As shown in the previous figures, $f_-(v^*; v_-)$ is monotone increasing, and $f_+(v^*; v_+)$ decreasing. Both are continuous, and indeed in $C^2(\mathbb{R}^+)$, even at the point $v_\pm$ where the wave changes from a shock to a rarefaction wave. Moreover, we may find the limits

$$\lim_{v^*\to 0+} f_-(v^*;v_-) = -\infty, \quad \lim_{v^*\to +\infty} f_-(v^*;v_-) = \frac{2\sqrt{\gamma}}{\gamma-1}v_-^{(1-\gamma)/2},$$
$$\lim_{v^*\to 0+} f_+(v^*;v_+) = +\infty, \quad \lim_{v^*\to +\infty} f_+(v^*;v_+) = -\frac{2\sqrt{\gamma}}{\gamma-1}v_+^{(1-\gamma)/2}. \tag{4.109}$$

Therefore, these two curves insect if and only if

$$u_+ - u_- < \frac{2\sqrt{\gamma}}{\gamma-1}\left(v_-^{(1-\gamma)/2} + v_+^{(1-\gamma)/2}\right). \tag{4.110}$$

Given (v_-, u_-), we may draw two curves, namely $u^* = u_- + f_-(v^*;v_-)$ and $u^* = u_- - f_+(v_-;v^*)$. They separate (v^*, u^*) plane into four subdomains, as shown in Fig. 4.17. Another curve $u^* = u_- + \frac{2\sqrt{\gamma}}{\gamma-1}\left(v_-^{(1-\gamma)/2} + v^{*(1-\gamma)/2}\right)$ takes out a region where no Riemann solution exists.

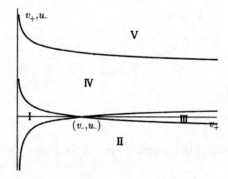

Figure 4.17 Riemann solutions for polytropic gas: state space.

For instance, if we have a Riemann problem with (v_+, u_+) in Region I, then we may draw a curve $u^* = u_+ + f_+(v^*;v_+)$, it intersects the curve $u^* = u_- + f_-(v^*;v_-)$ at a point (v^*, u^*). In fact, this is exactly the intermediate state we are looking for. Geometrically, $u^* = u_- + f_-(v^*;v_-)$ are the states (v^*, u^*), which may be reached from (v_-, u_-) through a backward wave; and $u^* = u_+ + f_+(v^*;v_+)$ are the states (v^*, u^*) which may reach (v_+, u_+) through a forward wave. Moreover, we observe that (v^*, u^*) lies in the shock branch of $u^* = u_- + f_-(v^*;v_-)$, and rarefaction branch of $u^* = u_+ + f_+(v^*;v_+)$. As a result, the Riemann problem is solved by a piecewise continuous solution with a backward shock and a forward rarefaction wave.

Other cases may be discussed in the same fashion. It is worth noticing that in the Region V, no intersection points may be found of the two curves, even when $v^* \to +\infty$. This corresponds to a situation of vacuum, for which our governing equations fail. Through rarefaction waves, u increases from left to right. This increment is bounded by v_- and v_+. The Region V is the place where the limitation of rarefaction wave occurs. See Fig. 4.18.

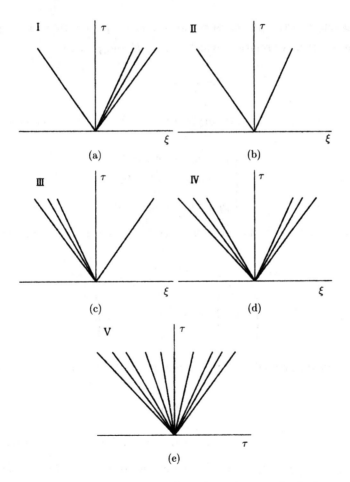

Figure 4.18   Riemann solutions for polytropic gas: physical space.

4.8 Elementary waves in a polytropic ideal gas

Physically speaking, entropy increases across a shock wave. Accordingly, the Euler equations in the previous section is not a realistic model for describing shock waves in gas flow. In this section, we consider a polytropic ideal gas as follows,

$$\begin{pmatrix} \rho \\ \rho u \\ E \end{pmatrix}_t + \begin{pmatrix} \rho u \\ \rho u^2 + p \\ (E+p)u \end{pmatrix}_x = 0. \tag{4.111}$$

These equations describe the conservation of mass, momentum and energy, respectively. The constitutive relation, which relates E and p, reads

$$E = \frac{p}{\gamma - 1} + \frac{1}{2}\rho u^2. \tag{4.112}$$

If we take ρ, u, p as the primary variables, the governing equations may be rewritten as

$$\begin{cases} \rho_t + u\rho_x + \rho u_x = 0, \\ u_t + uu_x + \dfrac{p_x}{\rho} = 0, \\ p_t + \gamma p u_x + u p_x = 0. \end{cases} \tag{4.113}$$

The corresponding eigenvalue problem takes eigenvalues and eigenvectors as follows,

$$\lambda^1 = u - c, \qquad \lambda^2 = u, \qquad \lambda^3 = u + c,$$

$$\mathbf{r}^1 = \begin{pmatrix} -\dfrac{\rho}{c} \\ 1 \\ -\rho c \end{pmatrix}, \mathbf{r}^2 = \begin{pmatrix} 1 \\ 0 \\ 0 \end{pmatrix}, \mathbf{r}^3 = \begin{pmatrix} \dfrac{\rho}{c} \\ 1 \\ \rho c \end{pmatrix}. \tag{4.114}$$

For the second eigen pair, we notice that when $(d\rho, du, dp)^T \parallel \mathbf{r}^2$, corresponding change in λ is

$$d\lambda \sim \nabla_{(\rho, u, p)} \lambda \cdot \mathbf{r}^2 = 0.$$

This gives rise to a new type of waves, called a slip line or a contact discontinuity. Across a slip line $\dfrac{dx}{dt} = u$, we have u, p unchanged, whereas an arbitrary jump in ρ may occur.

As a matter of fact, consider a general hyperbolic system

$$U_t + (F(U))_x = 0. \tag{4.115}$$

If for an eigen pair $(\lambda, \mathbf{r})$, it holds that $\nabla_U \lambda \cdot \mathbf{r} = 0$, we say that the corresponding field is linearly degenerate. In contrast, if it always holds that $\nabla_U \lambda \cdot \mathbf{r} \neq 0$, we call this a genuinely nonlinear field. The former field gives rise to contact discontinuity, and the latter gives shock or rarefaction wave. It may be readily shown that the other two eigen pairs in the polytropic ideal gas are genuinely nonlinear.

Now we describe the rarefaction waves for the third family. Same as before, we consider the self-similarity variable $\eta = x/t$ and a self-similar solution solves

$$\begin{cases} -\eta \rho' + u\rho' + \rho u' = 0, \\ -\eta u' + uu' + \dfrac{1}{\rho} p' = 0, \\ -\eta p' + \gamma p u' + u p' = 0. \end{cases} \tag{4.116}$$

By the Fredholm alternative, this linear system in (ρ', u', p') has nontrivial solution if and only if the coefficient matrix is degenerate, i.e.,

$$\begin{vmatrix} u-\eta & \rho & 0 \\ 0 & u-\eta & \dfrac{1}{\rho} \\ 0 & \gamma p & u-\eta \end{vmatrix} = (u-\eta)\left[(u-\eta)^2 - \dfrac{\gamma p}{\rho}\right]. \tag{4.117}$$

The root that corresponds to the third family is $\eta = u + \sqrt{\gamma p/\rho} = u + c$. The solution is $(\rho, u, p)^T \parallel \mathbf{r}^3$. Accordingly, we have

$$\frac{dp}{d\rho} = \frac{p'}{\rho'} = c^2 = \frac{\gamma p}{\rho},$$

which means $\dfrac{p}{\rho^\gamma} = \alpha$ is a constant. This actually gives a conserved quantity

$$s = c_v \ln \frac{p}{\rho^\gamma}. \tag{4.118}$$

It is a specific entropy with c_v the specific heat under constant volume. As a matter of fact, from the equations (4.113), it is straightforward to find that

$$s_t + u s_x = c_v \left(\frac{p_t + u p_x}{p} - \gamma \frac{\rho_t + u \rho_x}{\rho}\right) = c_v(-u_x + u_x) = 0.$$

For the rarefaction wave of the third family, s maintains constant. Moreover, we obtain consequently

$$\frac{\mathrm{d}u}{\mathrm{d}\rho} = \frac{c}{\rho} = \sqrt{\alpha\gamma}\rho^{\frac{\gamma-3}{2}}.$$

This gives another conserved quantity for centered rarefaction waves of the third family

$$u - \int \sqrt{\alpha\gamma}\rho^{\frac{\gamma-3}{2}}\,\mathrm{d}\rho = u - \frac{2c}{\gamma-1} = \beta. \tag{4.119}$$

In summary, the centered rarefaction solution of the third family is determined by the following three equalities,

$$u + c = \eta = \frac{x}{t}, \quad p = \alpha\rho^\gamma, \quad u - \frac{2c}{\gamma-1} = \beta. \tag{4.120}$$

For a shock wave of the third family with propagation speed v, we may take a vanishing viscosity approach to obtain the Rankine-Hugoniot relation

$$\begin{cases} -v[\rho] + [\rho u] = 0, \\ -v[\rho u] + [\rho u^2 + p] = 0, \\ -v[E] + [(E+p)u] = 0, \end{cases} \tag{4.121}$$

and the entropy condition

$$u_- + \sqrt{\frac{\gamma p_-}{\rho_-}} > u_+ + \sqrt{\frac{\gamma p_+}{\rho_+}}. \tag{4.122}$$

4.9 Soliton and inverse scattering transform

In this section, we discuss another type of waves, called solitary waves. It also arises from water waves, yet with dispersion. There does not develop any discontinuities.

In the year 1834, a young engineer John Scott Russell (1808-1882) observed, when he rode on the horseback along a canal in Scotland, a big wave propagating in a constant speed, in a persistent shape, for a very long time. He described in "Report on Waves" (1844) as follows.

I was observing the motion of a boat which was rapidly drawn along a narrow channel by a pair of horses, when the boat suddenly stopped - not so the mass of water in the channel which it had put in motion; it accumulated round the prow of the vessel in a state of violent agitation, then suddenly leaving it behind, rolled forward with great velocity, assuming the form of a large solitary elevation, a rounded, smooth and well-defined heap of water, which continued its course along the channel apparently without change of form or diminution of speed. I followed it on horseback, and overtook it still rolling on at a rate of some eight or nine miles an hour, preserving its original figure some thirty feet long and a foot to a foot and a half in height. Its height gradually diminished, and after a chase of one or two miles I lost it in the windings of the channel. Such, in the month of August 1834, was my first chance interview with that singular and beautiful phenomenon which I have called the Wave of Translation.

In 1895, Korteweg and de Vries published a paper "*On the change of form of long waves advancing in a rectangular canal and on a new type of long solitary waves*". In this paper, they derived a simple model, which bears their name as the KdV equation. Let u be elevation of the water, it solves

$$u_t + 6uu_x + u_{xxx} = 0. \tag{4.123}$$

We notice that the third order term u_{xxx} functions in a similar way as u_{xx} in Burgers' equation, namely, to prevent the formation of discontinuities. However, it differs by being *dispersive*, instead of *dissipative*. More precisely, let us consider a bounded interval $[-M, M]$, where we suppose that M is so big that there is essentially nothing outside of the interval, namely, $u(x,t)$ is close to 0 there. Consider Burgers' equation

$$u_t + uu_x - u_{xx} = 0. \tag{4.124}$$

We integrate over the interval,

$$\frac{\mathrm{d}\int_{-M}^{M} u\,\mathrm{d}x}{\mathrm{d}t} + \left[\frac{u^2}{2} - u_x\right]_{-M}^{M} = 0. \tag{4.125}$$

The difference of u^2 at the boundary might be neglected, and we may even translate the axis to make it exact. However, the u_x term is typically with different sign at the boundary points $-M$ and M, since u decreases to 0 at infinity. This makes a permanent leakage of the "total mass" $\int_{-M}^{M} u \mathrm{d}x$.

On the other hand, with the same treatment on the KdV equation, we obtain

$$\frac{\mathrm{d}\int_{-M}^{M} u \mathrm{d}x}{\mathrm{d}t} + \left[3u^2 + u_{xx}\right]_{-M}^{M} = 0. \tag{4.126}$$

This u_{xx} term, on the other hand, can be a gain or a loss of the total mass, and the difference at the boundary points over a long time period is negligible.

The difference between the dispersive and dissipative mechanisms may also be explained in terms of linear dispersion relations. With a equilibrium ground state $U(x,t) = 0$, the dispersion relations for Burgers' equation and the KdV equation are $\lambda = -\omega^2$ and $\lambda = i\omega^3$, respectively. While the first one represents a decreasing amplitude for each mode, the second one corresponds to an oscillation with varying frequencies for different wave modes.

With the KdV equation, we are ready to describe this solitary wave through traveling wave analysis. Named as a solitary wave, it is a single wave, maintaining amplitude and speed.

A traveling wave with speed c solves, with $\xi = x - ct$,

$$u''' + (6u - c)u' = 0. \tag{4.127}$$

Integrated once, it gives

$$u'' + 3u^2 - cu = C_1. \tag{4.128}$$

The wave we are looking for has the property $\lim_{x \to \infty} u(x,t) = 0$, therefore $C_1 = 0$. Now we consider the second order ODE

$$u'' + 3u^2 - cu = 0, \quad u(\pm\infty) = 0. \tag{4.129}$$

It can be integrated once again, after multiplied by u',

$$(u')^2 + 2u^3 - cu^2 = C_2. \tag{4.130}$$

For the same reason, C_2 has to be zero. The solution to the first order nonlinear ODE

$$(u')^2 + 2u^3 - cu^2 = 0 \tag{4.131}$$

is

$$u = \frac{a^2}{2}\text{sech}^2\left(\frac{a\xi}{2}\right), \quad a = \sqrt{c}. \qquad (4.132)$$

Here the hyperbolic secant function is defined by

$$\text{sech} z = \frac{1}{\cosh z} = \frac{2}{e^z + e^{-z}}. \qquad (4.133)$$

This solution is quite interesting in that the height, which is $a^2/2$, and the characteristic width, which can be taken as $2/a$, are related to the wave speed by $a = \sqrt{c}$. A taller wave is thinner, and runs faster.

Without going to the details, we point out that periodic solutions exist when C_1 and C_2 are non-zero. These solutions are called cnoidal waves.

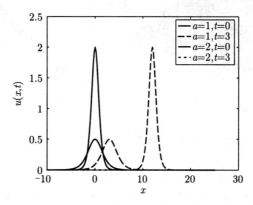

Figure 4.19 Two solitary waves with $a = 1$ and $a = 2$.

A solitary wave is a single wave, with both end-states $\lim_{x \to \pm\infty} u(x,t) = 0$. This type of wave cannot be supported by a dissipative system, such as Burgers' equation. There a traveling wave must take a shock profile, with $\lim_{x \to -\infty} u(x,t) > \lim_{x \to +\infty} u(x,t)$.

Mathematicians coined the word soliton, because they observed that the solitary waves in the KdV equation behave like a 'particle'. Note that elementary particles have names ending up with 'on', such as proton, electron, etc. The invention of soliton came from numerical results by Kruskal and Zabusky. They essentially put two solitary waves together, and numerically solved the KdV equation. Later on, it is found that they actually obtained a 2-soliton solution.

In general, a 2-soliton solution is

$$2\frac{\alpha_1^2 f_1 + \alpha_2^2 f_2 + 2(\alpha_2 - \alpha_1)^2 f_1 f_2 + [(\alpha_2 - \alpha_1)/(\alpha_2 + \alpha_1)]^2 \left(\alpha_2^2 f_1^2 f_2 + \alpha_1^2 f_2^2 f_1\right)}{\left[1 + f_1 + f_2 + [(\alpha_2 - \alpha_1)/(\alpha_2 + \alpha_1)]^2 f_1 f_2\right]^2},$$
(4.134)

where $f_i = e^{-\alpha_i(x - \alpha_i^2 t - x_i)}$.

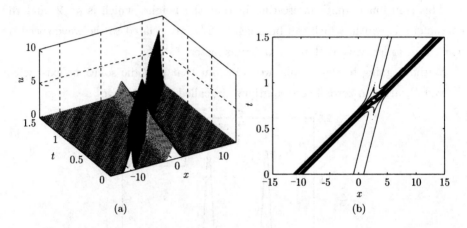

Figure 4.20 Two solitons interact in the KdV equation: (a) evolution; (b) contour lines.

One such example with $\alpha_1 = 2, \alpha_2 = 4$, and $x_1 = 0, x_2 = -10$ is depicted in Fig. 4.20. Initially we observe two solitary waves, with the taller one behind the shorter one. They propagate at different speeds, therefore the taller one catches up at a certain time. A complex combination of them occurs during this interaction period. Yet eventually, the taller one goes across, and leaves the shorter one behind. The most fascinating phenomenon is of course the "superposition" of two solitary waves. How can such a nonlinear system still possesses superposition after a nonlinear interaction? As a matter of fact, the interaction is not completely particle-like. A careful check shows that the phases of the two waves change after the crossing. This is better observed in the contour plot. By phase, we mean the place where each wave assumes its height. This shows the effect of nonlinearity.

If we replace $u(x,t)$ by $-u(x,t)$, the KdV equation may be recast to

$$u_t - 6uu_x + u_{xxx} = 0.$$
(4.135)

There is a remarkable way to solve this equation, namely, the inverse scattering transform (IST). IST was invented by Gardner, Greene, Kruskal and Miura (1969). Later on, an important generalization was made by Lax in terms of Lax pair.

We recall that the Cauchy problem for a linear system

$$u_t = u_{xx} + 3u \tag{4.136}$$

may be solved through the Fourier transform.

Step 1. For initial data $u(0,x) = u_0(x)$, we first make a Fourier transform to obtain its spectrum $U(0;\omega)$.

The governing equation after Fourier transform reads

$$U' = (-\omega^2 + 3)U. \tag{4.137}$$

Here ω is regarded as a parameter, and the evolution is decoupled for different modes.

Step 2. The solution in the spectral space is $U(t;\omega) = U(0,\omega)e^{(-\omega^2+3)t}$.

Step 3. We perform an inverse Fourier transform to obtain the solution at time t.

In a similar fashion, IST method is schematically shown in Fig. 4.21.

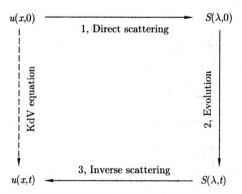

Figure 4.21 A diagram for the inverse scattering transform.

To illustrate this method, we recall the Schrödinger equation. It describes the evolution of a wave function $\psi(x,t)$ under a potential $V(x)$, which vanishes rapidly

at infinity,
$$i\hbar \partial_t \psi = -\frac{\hbar^2}{2m}\partial_{xx}\psi + V(x)\psi. \tag{4.138}$$

The Schrödinger equation is the fundamental equation in quantum mechanics, and therefore well studied during 1926 to 1960's. This provides a solid basis for the development of IST. The Schrödinger equation is a linear equation, thus people investigate the corresponding "eigenvalue" problem, after rescaling,
$$\lambda \psi = -\partial_{xx}\psi + V(x)\psi. \tag{4.139}$$

In fact, the eigenvalue λ stands for the energy level for the corresponding eigenfunction. More precisely, we call this a spectral problem, because ψ is a function instead of a column vector. In a linear PDE system such as the Schrödinger equation, there exist not only discrete eigenvalues, but also continuous spectra. In a direct scattering, one obtains the spectral data which includes the eigenvalues and eigenfunctions for a given potential $V(x)$. To illustrate the process, we consider a potential well
$$V(x) = \begin{cases} -H, & \text{if } |x| < 1, \\ 0, & \text{elsewhere.} \end{cases} \tag{4.140}$$

We first consider the case $\lambda = k^2 > 0$. The solution reads, in three intervals respectively,
$$\psi(x) = \begin{cases} A_1 e^{-ikx} + B_1 e^{ikx}, & \text{if } x < -1, \\ A_0 e^{-i\sqrt{k^2+H}x} + B_0 e^{i\sqrt{k^2+H}x}, & \text{if } -1 < x < 1, \\ A_2 e^{-ikx} + B_2 e^{ikx}, & \text{if } x > 1. \end{cases} \tag{4.141}$$

The C^1 continuity for the equation (4.139) at $x = \pm 1$ becomes
$$\begin{cases} A_1 e^{ik} + B_1 e^{-ik} = A_0 e^{i\sqrt{k^2+H}} + B_0 e^{-i\sqrt{k^2+H}}, \\ ik(-A_1 e^{ik} + B_1 e^{-ik}) = i\sqrt{k^2+H}(-A_0 e^{i\sqrt{k^2+H}} + B_0 e^{-i\sqrt{k^2+H}}), \\ A_2 e^{-ik} + B_2 e^{ik} = A_0 e^{-i\sqrt{k^2+H}} + B_0 e^{i\sqrt{k^2+H}}, \\ ik(-A_2 e^{-ik} + B_2 e^{ik}) = i\sqrt{k^2+H}(-A_0 e^{-i\sqrt{k^2+H}} + B_0 e^{i\sqrt{k^2+H}}). \end{cases} \tag{4.142}$$

With 4 linear equations for 6 unknowns, it may be readily checked that this linear system has infinite many solutions. To confine ourselves to a meaningful solution, we notice that actually the wave form for $x < -1$ is $e^{-i\lambda t}(A_1 e^{-ikx} + B_1 e^{ikx})$, which represents wave components in two directions. Therefore, we require $B_1 = 1$ as an incident wave from the left, and denote the reflection amplitude as $r(k) = A_1$. For $x > 1$, we only consider the transmitted wave with amplitude denoted as $a(k) = B_2$ whereas $A_2 = 0$. Then we end up with 4 unknowns which are uniquely solved from the linear equations.

Next, we consider the case $\lambda = -k^2 \in (-H, 0)$. The solution then reads

$$\psi(x) = \begin{cases} A_1 e^{-kx} + B_1 e^{kx}, & \text{if } x < -1, \\ A_0 e^{-i\sqrt{H-k^2}x} + B_0 e^{i\sqrt{H-k^2}x}, & \text{if } -1 < x < 1, \\ A_2 e^{-kx} + B_2 e^{kx}, & \text{if } x > 1. \end{cases} \quad (4.143)$$

Because $\psi(x)$ must keep finite due to the unitary property $\int_{\mathbb{R}} |\psi(x)|^2 dx = 1$, it holds that $A_1 = B_2 = 0$.

The C^1 continuity gives

$$\begin{cases} B_1 e^{-k} = A_0 e^{i\sqrt{H-k^2}} + B_0 e^{-i\sqrt{H-k^2}}, \\ kB_1 e^{-k} = i\sqrt{H-k^2}(-A_0 e^{i\sqrt{H-k^2}} + B_0 e^{-i\sqrt{H-k^2}}), \\ A_2 e^{-k} = A_0 e^{-i\sqrt{H-k^2}} + B_0 e^{i\sqrt{H-k^2}}, \\ -kA_2 e^{-k} = i\sqrt{H-k^2}(-A_0 e^{-i\sqrt{H-k^2}} + B_0 e^{i\sqrt{H-k^2}}). \end{cases} \quad (4.144)$$

Eliminating B_1 from the first two equations and A_2 from the last two, we obtain

$$\begin{cases} -(i\sqrt{H-k^2}+k)e^{i\sqrt{H-k^2}}A_0 + (i\sqrt{H-k^2}-k)e^{-i\sqrt{H-k^2}}B_0 = 0, \\ -(i\sqrt{H-k^2}-k)e^{-i\sqrt{H-k^2}}A_0 + (i\sqrt{H-k^2}+k)e^{i\sqrt{H-k^2}}B_0 = 0. \end{cases} \quad (4.145)$$

The solvability condition, by the Fredholm alternative, then reads

$$-(i\sqrt{H-k^2}+k)^2 e^{2i\sqrt{H-k^2}} + (i\sqrt{H-k^2}-k)^2 e^{-2i\sqrt{H-k^2}} = 0. \quad (4.146)$$

Or, equivalently,

$$\sqrt{H-k^2}\tan\sqrt{H-k^2} = k, \quad \text{or} \quad \sqrt{H-k^2}\cot\sqrt{H-k^2} = -k. \quad (4.147)$$

These algebraic equations have finite many roots, leading to discrete energy levels denoted as $k_n, n = 1, \cdots, N$. The eigenfunction corresponding to c_n, after rescaling under $\int_{\mathbb{R}} |\psi(x)|^2 dx = 1$ may be identified by $c_n \equiv B_2$.

On the other hand, if $\lambda = -k^2 < -H$, then all above discussions are valid, except that we need to replace $\sqrt{H-k^2}$ by $i\sqrt{H+k^2}$. The solvability condition (4.146) now reads

$$-(-\sqrt{H+k^2}+k)^2 e^{-2\sqrt{H+k^2}} + (\sqrt{H+k^2}+k)^2 e^{2\sqrt{H+k^2}} = 0. \quad (4.148)$$

This algebraic equation has no real root.

In summary, the eigenvalue problem for the Schrödinger equation has solution for all $\lambda > 0$ and no solution for $\lambda < -H$. In between, there are finite many eigenvalues. We collect the information $(\{k_n, c_n\}_{n=1}^N; r(k), a(k))$ and call it the scattering data for the potential $V(x)$.

It is interesting that actually the potential $V(x)$ is characterized by the spectral data in a unique way. In another word, given the spectral data, one may find the potential.

On the other hand, if we set $V(x;t) = u(x,t)$, the corresponding spectral data has a simple evolution rule, which we shall partially illustrate later on.

Making use of all above discoveries, we now present the inverse scattering transform (IST) consisting the following three steps.

Step 1. In the Schrödinger equation, we put the initial data of KdV equation $u(x,0)$ as the potential function $V(x)$, and extract the scattering data as follows,

$$S(0) = \left(\{k_n, c_n(0)\}_{n=1}^N; r(k;0), a(k;0)\right). \quad (4.149)$$

Here, $\{k_n, c_n\}$ is a pair for negative discrete eigenvalue $\lambda = -k_n^2$, with normalizing coefficient to make the eigenfunction $\psi_n(x) \sim c_n e^{-k_n x}$, $x \to +\infty$, and $\int_{\mathbb{R}} \psi_n^2(x) dx = 1$. The continuous part, on the other hand, describes positive eigenvalue $\lambda = k^2$ for $k \in \mathbb{R}^+$, which has eigenfunction $\psi(x)$ satisfying

$$\psi(x) \sim e^{-ikx} + r(k)e^{ikx}, \quad x \to +\infty; \quad \psi(x) \sim a(k)e^{-ikx}, \quad x \to -\infty. \quad (4.150)$$

Step 2. The scattering data evolves in the following manner,

$$S(t) = \left(\{k_n, c_n(t)\}_{n=1}^N; r(k;t), a(k;t)\right), \quad (4.151)$$

with

$$c_n(t) = c_n(0)e^{4k_n^3 t}, \quad r(k;t) = r(k;0)e^{8ik^3 t}, \quad a(k;t) = a(k,0). \quad (4.152)$$

Here the initial data are the scattering data obtained in Step 1.

Step 3. We recover the solution $u(x,t)$ from the spectral problem

$$\lambda \psi = -\partial_{xx}\psi + u(x,t)\psi. \tag{4.153}$$

That is, knowing the scattering data $S(t)$, we find the potential $u(x,t)$. We make two remarks here. First, the scattering data itself is not the complete information about the spectral problem, namely, we do not have the exact form of the eigenfunctions. Instead, we only know their asymptotes. Secondly, the inverse scattering step is done at each fixed time. That is, in this step, there is no time evolution. This is purely an ODE problem. Fortunately, in the 1960's, people knew well how to do this step. In fact, we define a function

$$F(x;t) = \sum_{i=1}^{N} c_n^2 e^{-k_n x} + \frac{1}{2\pi}\int_x^{+\infty} r(k;t)e^{ikx}\,dx. \tag{4.154}$$

Then we consider the solution to the Gel'fand-Levitan-Marchenko equation

$$K(x,y;t) + F(x+y;t) + \int_x^{+\infty} K(x,z;t)F(z+y;t)\,dz = 0. \tag{4.155}$$

The potential function is recovered from

$$u(x,t) = 2\partial_x\left[K(x,x;t)\right]. \tag{4.156}$$

Now we start to explain the motivation of IST. Of course this is *ad hoc*.

The key point in IST is to relate KdV equation with the spectral problem for the Schrödinger equation. This is initiated by the Miura transform

$$u = v^2 + v_x. \tag{4.157}$$

It leads to the mKdV (modified KdV) equation

$$v_t - 6v^2 v_x + v_{xxx} = 0. \tag{4.158}$$

In fact, we observe that $u_t = 2vv_t + v_{xt}$, $u_x = 2vv_x + v_{xx}$, $u_{xxx} = 6v_x v_{xx} + 2vv_{xxx} + v_{xxxx}$. The KdV equation then becomes

$$2vv_t + v_{xt} - 6(v^2 + v_x)(2vv_x + v_{xx}) + 6vv_{xx} + 2v_x v_{xxx} + v_{xxxx} = 0, \tag{4.159}$$

or,
$$(\partial_x + 2v)(v_t - 6v^2 v_x + v_{xxx}) = 0. \tag{4.160}$$

The mKdV equation is no easier than the KdV equation itself. However, the importance is not the mKdV equation. Instead, the transform inspires a well-known change of variable, because it is a Riccati equation on v.

In fact, if we replace the Miura transform by the GGKM transform
$$u = v_x + v^2 + \lambda, \tag{4.161}$$

and further let
$$v = \partial_x(\ln \psi) = \frac{\psi_x}{\psi}, \tag{4.162}$$

we end up with the eigenvalue problem for the Schrödinger equation
$$-\psi_{xx} + u\psi = \lambda \psi. \tag{4.163}$$

Another question is about the evolution of the scattering data. Here we discuss only for the discrete eigenvalues $\lambda = -k_n^2$. Noticing that the eigenvalue problem may be solved in real number space, we have the normalization condition $\int_{\mathbb{R}} \psi^2 \, dx = 1$.

Because $u = \psi_{xx}/\psi + \lambda$, we compute the terms in the KdV equation as follows,

$$u_t = \frac{\psi_{xxt}}{\psi} - \frac{\psi_t \psi_{xx}}{\psi^2} + \lambda_t,$$

$$u_x = \frac{\psi_{xxx}}{\psi} - \frac{\psi_x \psi_{xx}}{\psi^2},$$

$$u_{xx} = \frac{\psi_{xxxx}}{\psi} - \frac{2\psi_x \psi_{xxx}}{\psi^2} - \frac{\psi_{xx}^2}{\psi^2} + \frac{2\psi_x^2 \psi_{xx}}{\psi^3},$$

$$u_{xxx} = \frac{\psi_{xxxxx}}{\psi} - \frac{3\psi_x \psi_{xxxx}}{\psi^2} - \frac{4\psi_{xx} \psi_{xxx}}{\psi^2} + \frac{6\psi_x^2 \psi_{xxx}}{\psi^3} + \frac{6\psi_x \psi_{xx}^2}{\psi^3} - \frac{6\psi_x^3 \psi_{xx}}{\psi^4},$$

$$uu_x = \frac{\psi_{xx} \psi_{xxx}}{\psi^2} - \frac{\psi_x \psi_{xx}^2}{\psi^3} + \lambda \frac{\psi_{xxx}}{\psi} - \lambda \frac{\psi_x \psi_{xx}}{\psi^2}.$$

We then calculate

$$\psi^2(u_t - 6uu_x + u_{xxxx})$$
$$= \psi^2 \lambda_t + [\psi\psi_{xxt} - \psi_t\psi_{xx}] - 6\lambda[\psi\psi_{xxx} - \psi_x\psi_{xx}] + \psi\psi_{xxxxx} - 10\psi_{xx}\psi_{xxx}$$
$$-3\psi_x\psi_{xxxx} + 12\frac{\psi_x^2\psi_{xxx}}{\psi} - \frac{6\psi_x^3\psi_{xx}}{\psi^2}$$
$$= \psi^2\lambda_t + \partial_x\Big[(\psi\psi_{tx} - \psi_t\psi_x) + (\psi\psi_{xxxx} - \psi_x\psi_{xxx}) - 6\lambda(\psi\psi_{xx} - \psi_x^2)$$
$$-3(\psi_x\psi_{xxx} + \psi_{xx}^2 - \frac{2\psi_x^2\psi_{xx}}{\psi})\Big]$$
$$= \psi^2\lambda_t + \partial_x\left[\psi R_x - \psi_x R\right],$$

with
$$R = \psi_t + \psi_{xxx} - 6\lambda\psi_x - 3\frac{\psi_x\psi_{xx}}{\psi}.$$

Using the normalization condition, we integrate over $x \in \mathbb{R}$ to get $\lambda_t = 0$. Consequently, we have
$$\psi R_{xx} - \psi_{xx}R = 0. \tag{4.164}$$

With the Schrödinger equation, this is equivalent to
$$-R_{xx} + uR = \lambda R. \tag{4.165}$$

That is, R is an eigenfunction.

On the other hand, we notice that for the eigenfunction ψ, we may construct $\psi_2 = \psi\int\frac{\mathrm{d}x}{\psi^2}$ which is another eigenfunction. In fact, we check
$$\psi_{2,xx} = \psi_{xx}\int\frac{\mathrm{d}x}{\psi^2} + 2\psi_x\frac{1}{\psi^2} + \psi\left(\frac{1}{\psi^2}\right)_x = \frac{\psi_{xx}\psi_2}{\psi} = (u - \lambda)\psi_2. \tag{4.166}$$

Furthermore, the linear independency for ψ_2 and ψ is readily shown. The linear equation $-\Psi_{xx} + u\Psi = \lambda\Psi$ then admits general solution in the form of a linear combination of ψ and ψ_2. In particular, we have
$$R = C\psi + D\psi_2. \tag{4.167}$$

Now consider the asymptotic behavior at $+\infty$,
$$\psi \sim c_n e^{-k_n x}, \quad x \to +\infty. \tag{4.168}$$

This means $\psi_2 \sim c * e^{k_n x}$, $x \to +\infty$. From the definition, R vanishes asymptotically, hence $D = 0$.

Now we have

$$R = \psi_t + \psi_{xxx} - 6\lambda\psi_x - 3\frac{\psi_x\psi_{xx}}{\psi} = C\psi. \tag{4.169}$$

Multiplying by ψ, we obtain

$$(\psi^2/2)_t + \left(\psi\psi_{xx} - 2\psi_x^2 - 3\lambda\psi^2\right)_x = C\psi^2. \tag{4.170}$$

We integrate over $x \in \mathbb{R}$. The left hand side vanishes, because $\int_{\mathbb{R}} \psi^2 dx = 1$. Therefore, we have $C = 0$ as well. Now plugging the form $\psi \sim c_n e^{-k_n x}$, $x \to +\infty$ into the equation, we notice $u \to 0$ and have

$$c_{n,t}\psi^2/c_n + \left[(-k_n^2 - 3\lambda)\psi^2\right]_x = 0. \tag{4.171}$$

We recall that $\lambda = -k_n^2$. Moreover, from $\psi^2 \sim c_n^2 e^{-2k_n x}$, $x \to +\infty$, we deduce $(\psi^2)_x \sim -2k_n\psi^2$, $x \to +\infty$. The above equation becomes

$$c_{n,t}/c_n + 2k_n^2(-2k_n) = 0. \tag{4.172}$$

The evolution of c_n is then determined by $c_n(t) = c_n(0)\exp(4k_n^3 t)$.

A special case with IST method is the N-soliton solution $u(x,t) = -2\partial_{xx}[\ln\det(I+C)]$. Here I is the identity matrix, and $C = (c_{mn})$ with

$$c_{mn} = c_m(t)c_n(t)\frac{\exp\left[-(k_m+k_n)x\right]}{k_m+k_n}, \quad c_m(t) = c_n(0)\exp(4k_n^3 t) > 0. \tag{4.173}$$

Finally, we mention a special property of the KdV equation, that is, it possesses infinite many conservation laws. That is, there are infinite equalities for its solution in terms of

$$\partial_t T(u) + \partial_x X(u) = 0. \tag{4.174}$$

Here T is called the conserved density, and X the flux. A trivial conservation law is in the form of

$$\partial_t(\partial_x F) + \partial_x(-\partial_t F) = 0. \tag{4.175}$$

In the KdV equation, we let $u = w + \varepsilon w_x + \varepsilon^2 w^2$ (Gardner's transformation). The equation becomes, after some computations,

$$(1 + \varepsilon\partial_x + 2\varepsilon^2 w)\left[w_t - 6(w + \varepsilon^2 w^2)w_x + w_{xxx}\right] = 0. \tag{4.176}$$

Chapter 4 Hyperbolic Conservation Laws

Since ε is a parameter, and w is a smooth function, it may be shown that

$$w_t - 6(w + \varepsilon^2 w^2)w_x + w_{xxx} = 0. \quad (4.177)$$

This can be rewritten as the Gardner equation

$$w_t - (3w^2 - 2\varepsilon^2 w^3 + w_{xx})_x = 0. \quad (4.178)$$

Now we perform a perturbation type of expansion

$$w(x, t; \varepsilon) = w_0 + \varepsilon w_1 + \varepsilon^2 w_2 + \cdots. \quad (4.179)$$

Meanwhile, Gardner's transformation leads to

$$\begin{aligned} w &= u - (\varepsilon w_x + \varepsilon^2 w^2) \\ &= u - \left[\varepsilon \left[u - (\varepsilon w_x + \varepsilon^2 w^2)\right]_x + \varepsilon^2 \left[u - (\varepsilon w_x + \varepsilon^2 w^2)\right]^2\right] \\ &= u - \varepsilon u_x - \varepsilon^2 (u^2 - w_x) + \cdots. \end{aligned} \quad (4.180)$$

So, by equating coefficients of the same order of ε, we may express w_i as a function of u and its derivatives. The Gardner equation gives, by expansion, infinite conservation laws for w_i, hence for u. It turns out that at odd orders of ε, the conservation laws are trivial, and nontrivial at even orders. We list the first a few conservation laws as follows,

$$\begin{aligned} &u_t + (-3u^2 + u_{xx})_x = 0, \\ &(u^2)_t + (-4u^3 + 2uu_{xx} - u_x^2)_x = 0, \\ &(2u^3 + u_x^2)_t + (-9u^4 + 6u^2 u_{xx} - 12uu_x^2 + 2u_x u_{xxx} - u_{xx}^2)_x = 0. \end{aligned} \quad (4.181)$$

Assignments

1. What is the solution to

$$u_t + xu_x = 0, \quad (4.182)$$

with initial data $u(x, 0) = u_0(x)$? What about $u_t - xu_x = 0$?

2. Given a smooth monotone decreasing profile $u(x, 0) = u_0(x)$, can you find the smallest time t^* when there is a x^*, such that $\dfrac{\partial u}{\partial x}(x^*, t^*)$ becomes infinity in inviscid Burgers' equation?

3. For a scalar conservation law
$$u_t + (f(u))_x = 0, \tag{4.183}$$
with the underlying convection-diffusion system
$$u_t + f(u)_x = \varepsilon u_{xx}, \tag{4.184}$$
and $f(u)$ is convex (i.e. $f''(u)$ has a definite sign), show that a shock connecting two states u_- and u_+ must propagate at a speed $c = \dfrac{f(u_+) - f(u_-)}{u_+ - u_-}$. Is there any further condition on u_- and u_+?

4. Find the self-similar solution to a general conservation law
$$u_t + (f(u))_x = 0, \tag{4.185}$$
under the condition that $f(u)$ is convex.

5. Inviscid Burgers' equation may be rewritten as
$$(u^2)_t + (2u^3/3)_x = 0. \tag{4.186}$$
What are the corresponding shock solutions? How about rarefaction waves?

6. Discuss in detail one more case among (B),(C) and (D) for wave interaction in inviscid Burgers' equation.

7. Find the Rankine-Hugoniot relation and the entropy condition of a forward shock wave in a polytropic gas in the Eulerian coordinates, namely,
$$\begin{cases} \rho_t + (\rho u)_x = 0, \\ (\rho u)_t + (\rho u^2 + p(\rho))_x = 0. \end{cases} \tag{4.187}$$
Is this the same as that in the Lagrangian formulation?

8. For the above system, find the expressions for forward and backward rarefaction waves in the Eulerian formulation. Also find the Riemann variables if possible.

9. For the isentropic polytropic gas, find the second order derivatives of $f_+(v^*)$ and $f_-(v^*)$ at the point $v_\pm$, respectively.

10. Given (v_+, u_+), discuss profiles for the Riemann problems, depending on the value of (v_-, u_-).

11. Explore the case of interaction between a backward shock wave and a forward shock wave in the isentropic polytropic gas.
12. Find the dispersion relations for Burgers' equation and the KdV equation at $u = u_0$ respectively, and make comparison.
13. By phase plane analysis, show that periodic traveling waves exist for the KdV equation.
14. Verify the first three conservation laws for the KdV equation.

Index

adjoint operator, 68
ansatz, 66
attractive, 48

balance law, 53
bifurcation, 33
    bifurcation diagram, 35
    bifurcation point, 34
    controlling parameter, 34
    homoclinic bifurcation, 49
    Hopf bifurcation, 37
    period doubling, 44
    pitchfork bifurcation, 49
    saddle-node bifurcation, 35
    stable branch, 35
    stationary bifurcation, 37
    subcritical bifurcation, 49
    supercritical bifurcation, 36
    tangential bifurcation, 35
    transcritical bifurcation, 35
    unstable branch, 35
bilinear form, 89
blow-up, 34, 51, 60, 103

Cauchy sequence, 7
chaos, 38
characteristic line/curve, 95
comparison principle, 60, 88
completeness, 7
contraction, 4, 10
critical point, 2, 3, 14

center, 16
focus, 16
node, 15
saddle, 15
sink, 15
source, 15

Dirichlet boundary condition, 52
dispersion relation, 57, 130
domain of dependence, 116

equation
    Allen-Cahn equation, 70
    Belousov-Zhabotinskii equation, 84
    Buckley-Leverett equation, 99
    Burgers' equation, 72, 129
    competition equations, 63
    convection-diffusion equation, 73
    Duffing equation, 18, 23
    elliptic equation, 86
    Euler equations, 99
    evolutionary Duffing equation, 61, 75
    Fisher equation, 56
    Fitzhugh-Nagumo equations, 49, 56, 62
    Gardner equation, 141
    Gel'fand-Levitan-Marchenko equation, 137
    hyperbolic equation, 97
    inviscid Burgers' equation, 101

KdV equation, 129
Laplace equation, 50
linear advection equation, 95
Lorenz equations, 38
mKdV equation, 138
Navier-Stokes equations, vi, 72
partial differential equation (PDE), 50
Rayleigh-Benard equations, 39
reaction-diffusion equation, 53
Riccati equation, 138
Schrödinger equation, 134
shallow water equations, 100
traveling wave equation, 69
van der Pol equation, 26
wave equation, 69
ergodicity, 46

Feigenbaum number, 45
Fick's law, 54
fixed point, 3, 10
Fredholm's alternative, 67
functional, 10

genuinely nonlinear, 127
ground state, 57, 64

Hadamard's example, 51
Hamiltonian, 20
harmonic function, 83
hyperbolic conservation laws, 98

initial value problem (Cauchy problem), 51
instability
    structural instability, 20, 31

invariant domain, 60, 62
inverse scattering transform, 128
iteration
    Gauss-Seidel iteration, 47
    Jacobian iteration, 13
    Picard iteration, 3, 12

Lagrangian, 19
limit set, 29
linearly degenerate, 127
linearly independent, 5
logistic map, 38
Lyapunov function, 29

manifold
    stable manifold, 22, 40
    unstable manifold, 22, 40
minimization problem, 89
minimum principle, 88

Neumann boundary condition, 53
norm, 5
    L^2 norm, 7, 58
    maximum norm, 6
nullcline, 63

ODE, 1
    autonomous, 1
    non-autonomous, 1
    order of ODE, 1
operator, 9
orbit, 2
    heteroclinic orbit, 19, 40, 104, 117
    homoclinic orbit, 19, 24
    limit cycle, 24, 25
    semi-orbit, 2

perturbation method, 64
plane analysis, 18
polytropic gas, 126

range of influence, 116
regular point, 3
Riemann problem, 103, 121
Riemann variable, 115

separatrix, 22
slip line, 126
soliton, 128, 131
space, 4
    Banach space, 7
    closed subspace, 11
    function space, 6, 86
    Hilbert space, 68, 87
    normed space, 6
    nullspace, 68
    phase space, 2
    Sobolev space, 86
    vector space, 4
stability, 29, 30, 57
    L^2-stability, 58
    asymptotically stable, 31
    exchange of stability, 36
    exponentially stable, 31
state variable, 1
sub-solution, 61

super-solution, 61
superposition, iv

theorem
    Banach fixed point theorem, 10
    characterization theorem, 89
    Lagrange theorem, 33
    Lax-Milgram theorem, 90
    Poincare-Bedixson theorem, 25
trajectory, 2
transform
    Cole-Hopf transform, 73
    Gardner's transformation, 140
    Miura transform, 137

wave
    cnoidal wave, 131
    rarefaction wave, 104, 106, 111, 119, 122, 128
    shock, 101, 104, 106, 117, 128
        entropy condition, 105
        Lax entropy condition, 105, 119
        Rankine-Hugoniot relation, 105, 117
    solitary wave, 129, 131
    traveling wave, 69, 104, 130
weak derivative, 87
weak solution, 86, 96